Gems, granites, and gravels

GEMS, GRANITES, AND GRAVELS
Knowing and using rocks and minerals

R. V. DIETRICH
Central Michigan University

AND

BRIAN J. SKINNER
Yale University

Cambridge University Press
Cambridge
New York   Port Chester   Melbourne   Sydney

Published by the Press Syndicate of the University of Cambridge
The Pitt Building, Trumpington Street, Cambridge CB2 1RP
40 West 20th Street, New York, NY 10011, USA
10 Stamford Road, Oakleigh, Melbourne 3166, Australia

© Cambridge University Press 1990

First published 1990

Printed in the United States of America

Library of Congress Cataloging-in-Publication Data

Dietrich, Richard Vincent, *1924–*

Gems, granites, and gravels – knowing and using rocks and minerals / R. V. Dietrich and Brian J. Skinner.

p. cm.

ISBN 0-521-34444-1

1. Mineralogy. 2. Petrology. I. Skinner, Brian J., *1928–*
II. Title.
QE363.2.D52 1990
549 – dc20   90–1506

British Library Cataloguing-in-Publication Data

Dietrich, Richard V. (Richard Vincent), *1924–*

Gems, granites, and gravels.

1. Minerals. Rocks
I. Title   II. Skinner, Brian J. (Brian John), *1928–*
553

ISBN 0-521-34444-1 (hardback)

Contents

Preface		*page* vii
1	**The mineral world**	1
	Mineral: A definition	1
	Minerals on Earth	5
	Minerals in the universe	7
	Mineralogists and mineralogy	9
2	**Crystal realms**	13
	The crystalline state	13
	The study of crystals through the ages	15
	X-rays and crystal structure	23
	The physical properties of crystalline solids	25
3	**Mineral chemistry**	35
	Chemical elements and compounds	36
	Mineral formulas	43
	The naming of minerals	49
	The formation of minerals	51
	Mineral occurrences	53
4	**Rocks**	55
	Rock: A definition	56
	Rock components	57
	The classification and naming of rocks	61
	Rock origins	62
5	**Soils, dusts, and muds**	77
	Weathering	79
	Soils	81
	Movement of the regolith	85

Contents

6 Ores and ore minerals — 96
- Kinds of resources — 97
- Ore minerals — 99
- Metallic ore deposits — 101
- Nonmetallic ore deposits — 111
- Energy resources — 115
- The future — 116

7 Building materials — 120
- Building stones — 121
- Rock products — 130
- Some famous structures — 136

8 Rocks and minerals in diverse environments — 138
- Plate tectonics — 138
- The rock cycle — 147
- Epilogue — 149

Appendix 1 Chemical symbols and the periodic table — 151
Appendix 2 Identification of the common rock-forming minerals — 154
Appendix 3 The identification of rocks — 158
Index — 165

Preface

Minerals are everywhere around us. Gems in jewelry are minerals. Granites and most other rocks consist of mineral grains. Gravels, sands, and soils are mixtures of minerals. Even the dust in the air we breathe is made up of tiny mineral particles. But how often do we stop to consider the roles that minerals play in our everyday lives, their utility, or even their beauty?

Minerals, especially gems such as diamond and crystals of common minerals such as quartz, have fascinated people for thousands of years. Much of modern science can trace its roots to the studies of the early scientists who worked to decipher the chemistry and the physical properties of minerals. The atomic theory, inorganic chemistry, optics, crystal physics, and several other specialties in modern science evolved from the study of minerals.

Today, minerals keep us alive and ensure the continuity of our society. The houses we live in, the automobiles we drive, the roads we travel on, and almost everything else we touch are made of minerals or of materials derived from minerals. Indeed, our use of minerals and mineral products is so great that, on average, every person on Earth uses either directly or indirectly ten metric tons of minerals each year. In the United States and Canada and the countries of Western Europe, the amounts are higher – up to seventeen metric tons per year; in developing countries, such as Sri Lanka and Zaire, the consumption rates are lower. Even so, the fact remains: Everybody needs, uses, and depends on minerals, rocks, and soils and their products.

This book is an introduction to the exciting subject of mineralogy and related specialties such as petrology (the study of rocks), crystallography (the study of crystals), and soil science. We describe some of the great discoveries in mineralogy in a historical context. We attempt to tell the story of what minerals are, how they form, how they are distributed around the world, how we depend upon them, and where

Preface

one can go to see collections of beautiful specimens of both minerals and rocks.

We, the authors, have been privileged to spend our professional working lives involved in the study of minerals and rocks. We are grateful to our professors, our former and current colleagues, and our spouses for both their indulgence and their influence over the years; many of them have contributed in noteworthy ways to our knowledge and appreciation of minerals and rocks, and thus also to our joie de vivre. We hope that some of our enthusiasm for the subject will be transmitted to those who read this book.

1

The mineral world

MINERAL: A DEFINITION

Diamond: Nothing fires the imagination like a beautiful diamond (Plates 1 and 2). Stories surrounding this most romantic mineral punctuate history: Adventurers have lost their lives, queens their reputations, and kings their empires – all for the sake of the diamond.

One could be excused for thinking that a mineral as rare and valuable as diamond must be composed of extremely uncommon chemical elements. Two English chemists, Humphry Davy and Smithson Tennant, apparently thought so when they set out to determine the composition of diamond in the early 1800s. What they discovered, however, was that diamond is simply one form of a single, very common element: carbon. In fact, diamond has exactly the same chemical composition as the graphite in "lead" pencils and the charcoal used in barbecue grills.

More than a hundred years after the experiments of Davy and Tennant, during the early years of the 20th Century, it was finally discovered that the atoms of carbon in diamond and in graphite are packed together in different ways (Fig. 1.1). It is those different packings that give the two minerals such different appearances and properties.

After the differences in atomic packing in diamond and graphite were elucidated, another half century passed before scientists succeeded in making diamonds in the laboratory. By that time, the hypothesis had become accepted that natural diamonds must form within the earth at extremely high pressures, such as those that exist at depths below approximately 150 km (90 mi). Thus, it was no surprise when, in 1955, the General Electric Company scientists responsible for the first synthesis (Plate 3) found that very high pressures were required to force carbon atoms into the tight packing of diamond. Under lower-pressure conditions, graphite, with its looser atomic packing, forms. (The fact that diamond crystallites and polycrystalline films have been produced from

Gems, granites, and gravels

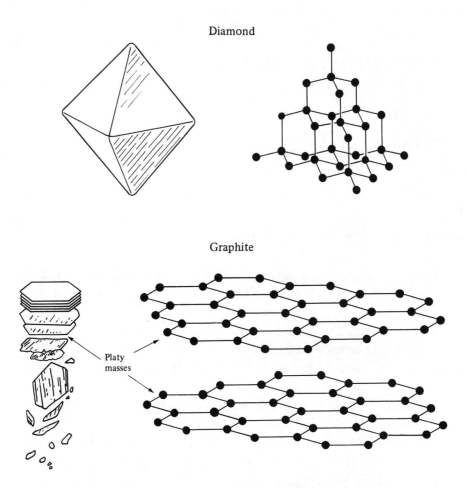

Figure 1.1. The packing of carbon atoms in diamond and graphite. In diamond, carbon atoms are linked in such a way that a very tough, three-dimensional network is formed. In graphite, the linkage of carbon atoms forms sheets. The sheets themselves are relatively tough, but they are held together by very weak bonds. Even gentle rubbing by the fingers is enough to overcome the intersheet forces, so graphite seems weak and is slippery.

carbon-rich vapors, e.g., sewer gas, at ambient pressure constitutes the proverbial "exception that proves the rule." This is so because the procedures used in such low-pressure syntheses involve processes that differ in several ways from those known or even thought possibly to exist in nature.)

Knowing that natural diamonds must have formed at such great depths raises a challenging question: How did diamonds reach the places on the Earth's surface where prospectors now find them? The answer is that they were carried upward by volcanism – not, however, by the kind of volcanism that occurs in the Hawaiian Islands or at Mount St. Helens.

Rather, it was an extraordinary volcanism, more highly explosive than has ever been observed by humans. In fact, the most recent episode of this kind of volcanism, generally called kimberlitic volcanism, appears to have occurred more than 50 million years ago, long before our ancestors appeared on Earth (Fig. 1.2).

The rarity of diamonds is the result of three factors:

1. As just implied, kimberlitic volcanism has occurred only infrequently during the Earth's long history.
2. Even when it has occurred, only tiny amounts of volcanic material have been carried up from great depths.
3. Perhaps most important, the proportion of diamonds in those host rocks is extremely small.

The richest deposits thus far discovered contain at most only 0.0000077 percent diamond, an amount equivalent to 0.2 g (= 1 carat) of diamond in three metric tons of rock – that is, an amount equivalent to a cube of diamond measuring approximately 0.4 cm (~5/32 in.) along an edge for every 20 wheelbarrow loads of crushed rock.

Three important points concerning minerals are illustrated by this brief introduction to diamonds:

1. Minerals are solids, not liquids or gases.
2. All minerals have definite chemical compositions; so all samples of a given mineral have the same composition. For example, when the chemical analysis of a diamond from South Africa is compared to that of a diamond from Russia, Australia, or anywhere else, they will always be found to have the same chemical composition.
3. The packing of the constituent atoms in all specimens of a given mineral is the same, no matter where the mineral is found.

An additional point, admittedly only implicit, is that specific minerals are formed under certain conditions of temperature, pressure, and chemical environment.

It follows from the fact that the atomic packing in a given mineral is always the same that the constituent atoms of any given mineral must have some regular geometric pattern. This is so because any given mineral grain contains billions of atoms, and it is highly improbable that any two random packings of billions of atoms would be identical. Indeed, the atoms in a mineral are never jumbled in a random, disorganized manner. Instead, they are packed in a repetitive, three-dimensional array that is called the mineral's *crystal structure*.

Because each mineral species does have a characteristic crystal structure, it follows that substances with the same composition but different

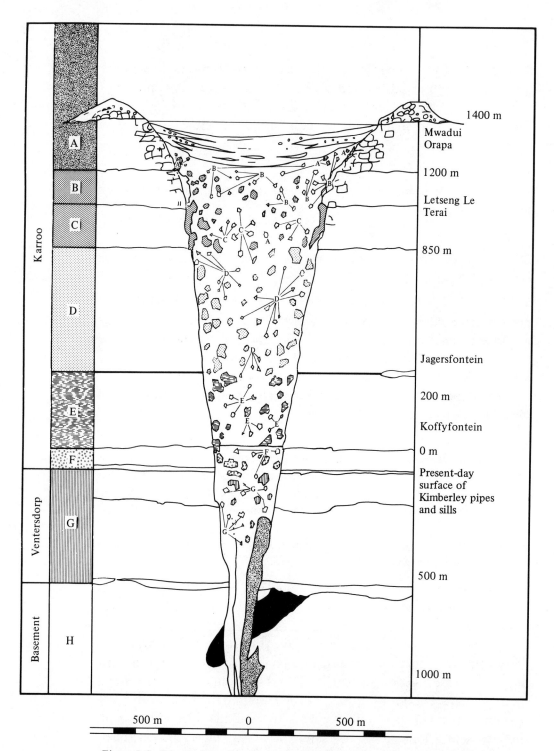

Figure 1.2. Diamonds are brought upward from depths of 150 km or more below the Earth's surface by the rare, explosive volcanic eruptions that form kimberlite pipes. This cross section of a pipe, shown with South Africa stratigraphic names (left column), depicts the typical flaring out near the surface and the way the pipe is filled with a rubble of smashed and broken rock fragments set in an uncommon igneous rock called kimberlite. Although humans have not witnessed the rare eruptions of a kimberlite, geological evidence suggests that it happens very

crystal structures – for example, diamond and graphite – must be different minerals. It also is true that minerals with the same crystal structure but different chemical compositions must be different minerals. Therefore, in order to be considered a valid mineral species, a mineral must have both a characteristic chemical composition and a characteristic crystal structure.

Thus, we come to a widely accepted definition for mineral:

A *mineral* is a naturally occurring inorganic solid with a specific chemical composition and a characteristic regular geometric arrangement of its constituent atoms.

Along with the already discussed aspects, this definition includes the words "naturally occurring" and "inorganic." "Naturally occurring" separates the small number of crystalline solids that occur in nature from the myriad solid compounds produced by chemists in their laboratories. "Inorganic" excludes the vast number of organic compounds that compose the cells of living plants and animals. It does not exclude minerals formed as a result of activities carried out by plants and animals – for example, the pearls and the shells of oysters, which are made up of mineral grains formed by the oysters.

MINERALS ON EARTH

Wherever we look around us, we see minerals. If we scoop up a handful of soil or sand, we will be holding a mixture of microscopic mineral grains (Plate 4). If we pick up a rock, again we most likely will be holding a mixture of minerals (Plate 5). This is true because soils, sands, and rocks are aggregates made up largely, if not wholly, of mineral grains. Soils and sands are loose aggregates; rocks are coherent aggregates.

The distinction between minerals and aggregates is important. When the word "mineral" is used, it should be applied only to an individual crystal or grain that strictly obeys the definition given for "mineral." On the other hand, most of the aggregates we call soil, sand, and rock are mixtures of many mineral grains, typically comprising two or more minerals.

Actually, minerals are only rarely found as single, pure entities. Instead, most of them occur as components of rocks, soils, and other natural

Caption to Figure 1.2 (*cont.*)
rapidly and explosively. (Diagram modified after an illustration by J. B. Hawthorne, in Wilson, A. N., 1982, Diamonds – From Here to Eternity, *by permission of Gemological Institute of America, Santa Monica, California)*

Gems, granites, and gravels

MINERAL SPECIMENS

Striking mineral specimens are among the Earth's most fascinating treasures. Each specimen is truly unique. Fine specimens have been sought, bought, traded, and displayed for centuries. Fortunately, the Earth continues to be a veritable storehouse of such specimens. For example, as new mines are opened, new areas prospected, and old mining areas reexamined, new mineral specimens are continually discovered. Thus, despite the millions of specimens already collected, fine specimens continue to be found every year. Even so, extremely fine specimens are exceedingly rare.

The collecting, trading, purchasing, and displaying of mineral specimens constitute a very special part of our culture. In fact, the collecting of minerals is one of the most popular of all hobbies, and mineral collections are major attractions in many of the world's museums.

The collecting of mineral specimens had become a rather sophisticated hobby by the 16th Century. Indeed, verified records indicate that several members of European royalty, as well as a few less prominent persons, had already accumulated fine "cabinets" full of specimens by the end of the 17th Century. Today, literally thousands of people have private collections, and some of those collections contain specimens that large museums would be proud to have as showpieces.

The collections of minerals in national, regional, and university museums are viewed by millions of people each year. A few museums — for example, the U.S. National Museum of Natural History (Smithsonian Institution), the American Museum of Natural History, the Royal Ontario Museum, the Vienna Naturhistorisches Museum, the British Museum of Natural History, and the Paris Muséum National d'Histoire Naturelle — have collections that number in the hundreds of thousands of specimens.

Each of the famous large collections is systematic, meaning that it contains specimens of a large number of the known mineral species arranged according to an accepted classification system. Each is also comprehensive in that it includes specimens that range from magnificent showpieces (Plate 6) to "reference specimens" for research. In addition, each includes special subcollections that are usually designated as such — for example, collections that have come as gifts and collections comprising groups of specimens that have been assembled in-house for special exhibits.

The collections of most universities and smaller museums and even of most private collectors are also systematic. A few, however, are rather specialized in that they are based on such things as

specimens of a single mineral or mineral group,
minerals and/or rocks from individual deposits or regions,
meteorites,
fluorescent minerals,
economic minerals,
rock-forming minerals and representative rocks,
rough and/or cut gem-quality minerals and/or rocks,
specimens to illustrate such things as physical properties or crystal morphology, and
micromounts, that is, specimens (typically fine crystals) that must be viewed by microscope to be appreciated.

Visit museum, university, and private mineral collections whenever the opportunity is presented, and be sure to allot sufficient time to really see, study, and appreciate the collections. Although it is demanding, gaining an appreciation of minerals is an exceptionally rewarding investment of time. Almost all collections – be they large or small, systematic or specialized – have at least a few specimens that are well worth seeing and studying. Furthermore, if one is considering the possibility of collecting minerals or rocks, it only stands to reason that the more collections viewed, the better one can decide what kind of collection might be the most fun or provide the greatest challenge. The diversity of possibilities is almost beyond belief; the only limit is the collector's imagination.

For a descriptive listing of Canadian and United States museums that have mineral exhibits, attention is directed to Matthew's compilation; for information about eighty of the more important mineral museums of Western Europe that are open to the public, see the Burchard and Bode book (both of these publications are listed in the "Further reading" section at the end of this chapter). Also, there are fine museums in many other countries – Australia and some of the countries in Asia, South America, and Africa. Many mineral exhibits are in museums or museum complexes that also have exhibits dealing with other subjects, such as paleontology, zoology, botany, archaeology and anthropology, art, history, or even antiques – something for nearly everyone.

---

aggregates, which are discussed in Chapters 4–6. This is not to say that specimens of individual minerals cannot be found or, if desired, separated from the aggregates in which they occur. Of course they can. Indeed, some of the most beautiful of nature's handiworks are minerals that were once part of mineral aggregates. One can see strikingly beautiful minerals on display in many of the world's most famous museums (Fig. 2.8 and Plates 1, 2, 6, 8, 13–16, 18, and 36–39).

MINERALS IN THE UNIVERSE

As a consequence of the engineering and scientific developments of the space age, astronauts have landed on the moon, and unmanned space vehicles have landed on and carried out experiments on Mars and Venus and have sent back close-up images of the surfaces of Mercury, the moons of Jupiter, Saturn, Neptune, and Uranus. Consequently, it is now known that each of those planetary objects is largely, if not wholly, solid and composed chiefly of minerals, that those extraterrestrial minerals are the same kinds of minerals as occur on Earth, and that the minerals typically occur in rocks and in loose aggregates very like those on Earth (Fig. 1.3).

Gems, granites, and gravels

*Figure 1.3. Similar rocks from the Earth and the moon. Both are igneous rocks called basalt; they contain the minerals olivine, pyroxene, and plagioclase feldspar. The holes are frozen bubbles produced by gas coming out of solution when the basaltic magma was erupted as a lava. (**a**) Vesicular basalt from Hawaii. (Photograph by B. J. Skinner) (**b**) Vesicular basalt from the moon; this specimen is seen here mounted for initial examinations by scientists at the Johnson Space Center, Houston. (Photograph courtesy of NASA)*

The discovery that other bodies in our solar system are made up of familiar minerals had been long expected, for several reasons. The most important reason was that for many years mineralogists had been studying meteorites, which are specimens from other parts of our solar system.

Meteorites are tiny planetary bodies that have fallen, or, more correctly, have been captured, when the Earth's gravitational field has pulled them from their former orbits around the sun. Meteorites are, then, of extraterrestrial origin; yet the minerals in them are the same as minerals in terrestrial rocks (Fig. 1.4). By extrapolation, it has been concluded that all natural inorganic solid objects in our solar system are made up of mixtures of minerals. That probably means that minerals occur here and there throughout the universe.

Granted, suns are so hot that they are entirely or largely gaseous, and thus there can be no minerals in suns. Granted also, it is suns that we see as twinkling "stars" when we look at the heavens at night. But in all likelihood, hard, rocky planets very much like the rocky planets of our own solar system are present around many of those suns, and those rocky planets surely consist, at least in part, of minerals.

Unfortunately, astronomers have not yet been able to observe planets outside our own solar system. That is true because planets, unlike suns, do not emit their own light; they only reflect light. Astronomers have,

The mineral world

Figure 1.4. Stony meteorite that fell at Weston, Connecticut, in 1807. The fluted surface is a crust formed by frictional heating as the meteorite fell through the Earth's atmosphere; the flat, cut surface reveals an internal assemblage of rocky fragments containing minerals such as olivine and pyroxene. Meteorites such as Weston are believed to have originated about 4.5 billion years ago as part of the formative processes of the solar system. (Photograph courtesy of Peabody Museum, Yale University)

however, observed minerals, in another form, well out in the universe: If one looks at the Milky Way on a clear night, it quickly becomes apparent that a dark cloud of some sort obscures the center of the Way. By careful measurements, astronomers have discovered that the cloud consists largely of fine, dust-sized mineral grains (Fig. 1.5).

In sum, on the basis of their observations scientists are certain that minerals occur here and there throughout our solar system. As a result of quite logical extrapolations, they are confident that minerals also occur here and there throughout the universe.

MINERALOGISTS AND MINERALOGY

The study of minerals is a specialized branch of science called *mineralogy*. Those educated in the subject are called *mineralogists*.

Gems, granites, and gravels

Figure 1.5. Dust "clouds" in the Milky Way. This view, which features the Way with its sporadic light-absorbing "clouds," was obtained with the superwide-angle camera of the Astronomisches Institut der Ruhr-Universität Bochum. The field is 140°, and the illumination is uniform (>93%) out to the edge. (Photograph by T. Schmidt-Kaler and W. Schlosser)

Although the study of minerals goes far back into antiquity, the modern science of mineralogy dates only from the early years of the 19th Century, when three lines of inquiry came together to create it. Those fields of inquiry were crystallography, the study of crystals; mineral chemistry, one of the antecedents of modern chemistry; and geology, the science of the earth, particularly as it involves the study of the rocks of which the Earth is built. These three parental disciplines are discussed briefly in the next three chapters.

FURTHER READING

Berry, L. G., and Mason, B., 1983, *Mineralogy: Concepts, Descriptions, Determinations* (2nd edition by R. V. Dietrich). W. H. Freeman, San Francisco, 561p.
This widely acclaimed introductory mineralogy text elaborates on both the definitions and the history given in this chapter.

Burchard, U. (with photographs by R. Bode), 1986, *Mineral Museums of Europe*. Walnut Hill Publishing Co. (no address given), 269p.
This book, with beautiful color illustrations, describes eighty of the more important mineralogical museums in Western Europe. Historical summaries and details about the exhibits – including lists of specimens deemed "excellent," "very good," "good," "remarkable," or "rarities" – are included.

Dietrich, R. V., and Chamberlain, S. C., 1989, Are cultured pearls mineral? – The continuing evolution of the definition of mineral. *Rocks & Minerals*, 64, 386–92.
This article discusses three somewhat controversial materials – biogenic minerals, mineraloids, and synthetic minerals – sometimes questioned as minerals. The authors support continuing evolution of the definition of mineral, along with evolution of mineralogy the science.

Dietrich, R. V., and Wicander, E. R., 1983, *Minerals, Rocks, and Fossils*. Wiley, New York, 212p.
This inexpensive paperback "self-teaching guide" is for those who may want to collect minerals, rocks, and/or fossils.

Dodd, R. T., 1986, *Thunderstones and Shooting Stars – The Meaning of Meteorites*. Harvard University Press, Cambridge, Massachusetts, 196p.
An up-to-date summary of what is known and thought about meteorites.

Frondel, J. W., 1975, *Lunar Mineralogy*. Wiley, New York, 323p.
A well-illustrated description of minerals found in samples collected by the Apollo 11 through Apollo 17 (U.S.) and Luna 16 and 20 (USSR) missions to the moon. Except for the absence of minerals with water content among the moon minerals, it is evident that the minerals found on the moon are essentially the same as those found on Earth.

McSween, H. Y., Jr., 1987, *Meteorites and Their Parent Planets*. Cambridge University Press, 237p.
A recently published, well-illustrated treatment of different kinds of meteorites and the kinds of bodies from which they are thought to have been derived.

Mason, B., and Melson, W. G., 1970, *The Lunar Rocks*. Wiley, New York, 179p.
This publication covers only the rocks known as a result of information gathered up through Apollo 12 (1969).

Matthews, W. H., III, 1977, *Mineral, Fossil & Rock Exhibits & Where to See Them*. American Geological Institute, Falls Church, Virginia, 56p.
This annotated list of "museums or other facilities in the United States and Canada that have exhibits or displays of geological interest" is rather difficult to use in that the information is coded to a tabulation (A through R) rather than being listed by institution. Data are now being updated for a new edition.

Medenbach, O., and Wilk, H., 1977, *The Magic of Minerals* (English translation by J. S. White, Jr.). Springer-Verlag, Berlin, 205p.
This coffee-table book has excellent color plates showing the kinds of quality mineral specimens that are present in fine museum and private collections.

Pough, F. H., 1976, *A Field Guide to Rocks and Minerals* (4th edition). Houghton-Mifflin, Boston, 317p.
A fine field guide for mineral collectors.

Robinson, A. L., 1986, Is diamond the new wonder material? *Science*, 234, 1074–6.
An update about ongoing research relating to optical and electronic applications – "diamond technology."

Titamgim, R. D., 1984, A bibliography of general mineral locality publications for the United States and Canada. *Rocks & Minerals*, 51, 203–9.
This bibliography lists types of coverage (inclusion of maps, etc.), sources, and prices of the available publications about mineral localities, many of which are still open to collectors.

Wilson, A. N., 1982, *Diamonds: From Birth to Eternity*. Gemological Institute of America, Santa Monica, California, 450p.
The illustrations alone make this rather rambling account worth scanning.

PERIODICAL

Rocks & Minerals. Heldref Foundation, 4000 Albemarle St., N.W., Washington, D.C. 20016. This bimonthly publication is directed primarily to amateur mineral, rock, and fossil collectors.

2

Crystal realms

THE CRYSTALLINE STATE

Anyone who has seen glistening needles of ice covering windows, bushes, or trees on a frosty morning (Plate 7) or sparkling stiletto-shaped icicles hanging from the eaves of a snow-covered roof can understand why the beauty of ice so fascinated the ancient Greeks and Romans. The Greeks called this beautiful transparent material *krystallos;* the Romans Latinized it to *crystallum*.

The Greeks and Romans were also fascinated by the water-clear, six-sided prisms of quartz (Plate 8) that came from the Alps. Those prisms also sparkled and glistened, much like icicles. Consequently, even though the prisms were not cold like ice, the idea developed that transparent quartz was a form of ice that apparently grew under intensely cold conditions, such as those that prevail in the high Alps. In fact, the idea became so ingrained in people's minds that by the Middle Ages transparent quartz was called "rock crystal," meaning, of course, rock-ice.

By the end of the Middle Ages, however, the belief in rock-ice was fading. Nonetheless, the word "crystal" remained in use, and eventually it came to include any shiny, transparent solid with flat, smooth faces. Hence, all flat-surfaced minerals became "crystals," and by the 17th Century "crystal" had attained its current meaning:

> A *crystal* is any solid object bounded by naturally flat, smooth surfaces.

During the 17th Century, crystals started to become objects of special study. Perhaps the most important observation made by scientists of the time was that the flat surfaces, called *crystal faces,* on minerals always appeared to be arranged in fixed, geometric ways. That was intriguing, especially for mathematicians. Geometry, it seemed, was not just an abstract idea dreamed up by philosophers; instead, considering the fact

Gems, granites, and gravels

(a)

(b)

Figure 2.1. The ways solids break. **(a)** *Calcite* ($CaCO_3$) *has a fixed internal order among its constituent atoms. Forces holding the atoms together are weaker in some directions than in others, and breakage occurs preferentially along the directions of weakness. As can be seen, there are three such directions in calcite. (Photograph courtesy of Ward's Natural Science Establishment)* **(b)** *Natural glass, such as that of which this tektite (see also Figure 4.9) is made, lacks internal order among its atoms. One consequence is that it breaks along curved, irregular fractures. This specimen, from the Khorat Plateau, Thailand, is from the collection of R. J. Lauf, Oak Ridge, Tennessee. (Photograph by R. V. Dietrich)*

that precise geometric forms grew spontaneously on minerals in rocks, geometry surely must have something to do with nature itself.

Near the end of the 17th Century, chemists also became aware of crystals. Their interest was piqued because they frequently saw masses of tiny crystals as they looked carefully at the solids that precipitated from their solutions.

As a result of these kinds of observations, the number of scientists who studied crystals steadily increased. In fact, before the end of the 18th Century, crystal studies had become a major activity for a rather large number of scientists.

One of the things that those early mathematicians, chemists, and other scientists did was to look at solids – both natural and artificially prepared – through microscopes. As they did, they found that tiny crystals and grains with flat surfaces are common, even in many seemingly featureless powders. Consequently, they concluded that

1. the capacity for crystals to form must depend on some kind of internal order (Fig. 2.1a), and
2. the ordered particles must be exceedingly small.

They also saw, however, that certain other solids – for example, glass – seemed never to grow flat faces. Whenever glass solidified, it tended to take on rounded shapes like those of a liquid; whenever glass was broken, it tended to break along irregular, curved surfaces rather than along the flat breakage surfaces exhibited by many minerals (Fig. 2.1b). Thus, they concluded that glass must lack any internal, geometrically regular order.

We now know that, just as those early scientists suspected, the atoms in glass are not organized in a strict geometric fashion; instead, these atoms have a random arrangement, like the atoms in liquids. Solids with such randomly jumbled atoms are said to be *amorphous*, from the Greek *a-*, meaning without, and *morphē*, meaning form.

Today, scientists distinguish two kinds of solid substances:

crystalline solids, which have internal order, and
amorphous solids, which lack internal order.

Minerals are, by definition, crystalline solids.

THE STUDY OF CRYSTALS THROUGH THE AGES

Consideration of crystals led people to ponder many questions: some ridiculous, such as whether or not crystals have mystical curative powers; others sublime, such as how many basic shapes can be found among crystals and what geometric principles are required to explain the shapes of all crystals. Questions such as the latter led early scientists to make extremely careful measurements of the external shapes of crystals, and those measurements eventually gave rise to the field of study we now call *crystallography*.

The underlying basis for distinguishing between amorphous and crystalline solids was the surmise that the spontaneous appearance of flat faces on crystals is an indication of some regular, precise internal structure. The nature of the constituent particles that made up such internal structures, however, remained a question throughout the 17th, 18th, and 19th Centuries, because atoms, as we know them today, were not recognized until early in the 20th Century. Consequently, all of the earlier attempts to address questions relating to the internal order of crystals were necessarily conjectural. Some of the conjectures, however, were remarkably perceptive, and by the time atoms were actually discovered, many of the principles governing the way they are packed within minerals had already been suggested.

Internal order in crystals

The first modern ideas dealing with the internal order in crystals appeared in the book *Micrographia*, written by Robert Hooke in 1665 (Fig. 2.2).

Gems, granites, and gravels

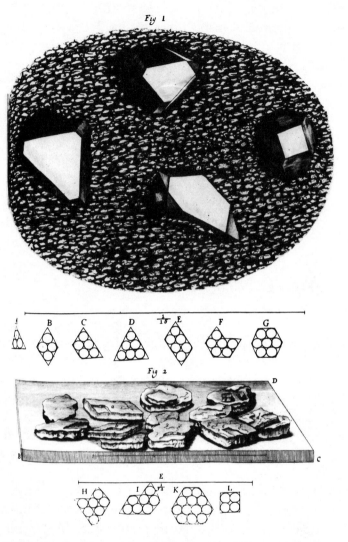

Figure 2.2. Plate V from Micrographia, *by Robert Hooke, published in London in 1665. The left side of the page appears distorted because one does not flatten one of the few existing copies of a rare classic such as this one – which is in the Beinecke Rare Book and Manuscript Library, Yale University – just to obtain a more accurate reproduction. When Hooke examined the broken surface of flint from Cornwall with a microscope, he observed tiny crystals, which he called "Cornish Diamants" (Fig. 1, upper sketch). They were probably tiny crystals of quartz. Hooke's demonstrations show how he thought the observed shapes of "Cornish Diamant" (Fig. 1A–G) and of alum crystals (Fig. 2H–L) could be due to regular packings of minute spherical particles.*

Hooke suggested that the ultimate particles in nature were tiny spherical bodies that were too small to be seen. Crystals, he said, must comprise regular stackings of those spheres, and to illustrate his point he stacked spherical cannonballs to demonstrate how several regular geometric solids could be so constituted. The real spheres, he suggested, were so small that the surfaces of the resulting solids appeared smooth even at high magnification.

The next step was taken by a Dutchman, Christiaan Huygens, who was a contemporary of Hooke. Huygens made many of the early important discoveries relating to the grinding of glass lenses for microscopes and telescopes. His interest in optics led him to study the passage of light through crystalline solids, as well as through glass. He found – as anyone who wishes can see today – that the properties of light passing through glass are the same in all directions, whereas the passage of light through crystals of many minerals is different in different directions.

Crystals of one particular mineral, then called Iceland spar, but today more commonly called calcite ($CaCO_3$), were the subject of particular study. Among other properties, Huygens noted that Iceland spar always broke along three nonparallel flat surfaces, so that the resulting fragments were rhombohedra. He suggested that the planar breaks (*cleavage directions,* as we now call them) must represent some natural direction of breakage between the stacked spheres envisioned by Hooke. Huygens thought, however, that any solid made up of spheres should break into cubes, not rhombohedra. Thus, he suggested that Iceland spar must consist of tiny ellipsoidal particles, rather than spheres.

Hooke and Huygens were both partly correct. The growth of crystal faces and the existence of cleavage directions do indeed arise from the internal stacking of atoms within crystals. But neither Huygens nor Hooke guessed, or apparently even imagined, how extremely tiny atoms actually are or how complicated the internal order within crystals can be.

During the next two centuries, many of the ideas about the internal order of crystals that were first expressed by Hooke and Huygens were refined. No major exceptions to their main observations, however, were voiced, and eventually, when the discovery and use of X-rays provided a way to actually measure the internal order within crystals, their ideas were proved correct. Before that time, though, several advances were made, with the crystallographers who were studying the external shapes of crystals leading the way.

The external shapes of crystals

The first important step in the study of the shapes of crystals was made by Niels Stensen, a Danish physician, better known by his Latinized name Nicolaus Steno. In 1669, Steno published the results of his studies of crystals of quartz (Fig. 2.3). The shapes of quartz crystals are highly variable because the relative sizes and shapes of their faces differ from

Gems, granites, and gravels

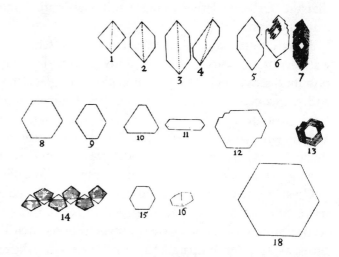

Figure 2.3. Part of a plate from De Solido, *by Nicolaus Steno, 1669. The figures reproduced here show how Steno illustrated his discovery that "the number and length of the sides [of crystals of a mineral (here quartz)] are changed in various ways without changing the angle" – that is, the law of constancy of interfacial angles.*

sample to sample. As Steno discovered, however, the angle between any given pair of faces is always the same. This important law – the *law of the constancy of interfacial angles* – is one of the basic tenets of crystallography. Steno explained the phenomenon by pointing out that in order for a crystal to grow, it must do so by the addition of new material, layer after layer, on an initial crystal nucleus. And although the thickness of a layer on a given face must be essentially constant, the thicknesses can differ considerably from face to face. Therefore, as a crystal grows, even though some faces may grow large relative to other faces, the angles between the faces will always remain the same.

Despite Steno's important work, two 18th-Century French mineralogists usually are considered to have been the fathers of crystallography: Jean Baptiste Louis Romé de l'Isle and the abbé René Just Haüy. Romé de l'Isle proved that crystals of the same substance – be they natural or synthetic – are identical. He also proved that, taken as a group, the angles between adjacent faces (the so-called *interfacial angles*) of a crystal are characteristic properties of a mineral – that is, minerals can be identified by measuring all of the interfacial angles. Haüy wrote several books about mineralogy and crystallography between 1784 and 1822 in which he presented evidence to prove that the constancy seen in interfacial angles is also true for cleavages – that is, the angles between cleavages are always the same for a given mineral. He also pointed out that interfacial angles seemed to have only certain values.

Those findings led Haüy to speculate about the way basic particles could be packed inside a crystal, and as a result he showed that within a crystal made by stacking objects of a fixed size and shape, faces could develop only in certain directions. That determination eventually led to another of the basic laws of crystallography: the *law of rational indices*. This law, which has been formulated in many different ways, states that when a set of crystal faces is described geometrically by referring the faces to three graphical axes of reference, the faces cut the axes only at simple multiples of the axial units. The implications of this law led both mineralogists and mathematicians to examine the geometry of crystal faces even more closely.

The crystal systems

Two German crystallographers, Christian Samuel Weiss and Friedrich Mohs, attacked the question of rational indices from a purely geometrical-mathematical viewpoint. They quickly found that three mutually perpendicular axes with equal units of measurement along each axis would describe the geometry of some crystals, but would give irrational results for many others. They then discovered that it was necessary only for the unit lengths to be different along the axes and/or for the axes to be at angles other than 90°. So the variables found to obtain were

1. the number of axes (three or four),
2. the angles between the axes, and
3. the unit distances along the axes.

These variables, used to define appropriate sets of axes, make the law of rational indices completely applicable.

Subsequent studies have shown that both the internal orders of atoms and the external geometries of crystal faces of all minerals – indeed, of all crystalline solids – can be described in terms of six different sets of geometric axes (Fig. 2.4). These sets of axes, now used to describe the *six crystal systems,* have been given the following names: cubic (also called isometric), tetragonal, hexagonal, orthorhombic, monoclinic, and triclinic.

The symmetry of crystals

While crystallographers were carrying out the studies that defined the six crystal systems, they noted that the faces on crystals could also be described by a system using features now called *elements of symmetry* (Fig. 2.5). For example, if all the faces of a crystal are arranged in parallel pairs on opposite sides of a point at the center of the crystal, it can be said that the crystal has a *center of symmetry*. A cube, for example, has a center of symmetry. Likewise, a crystal has an *axis of symmetry* if the

Gems, granites, and gravels

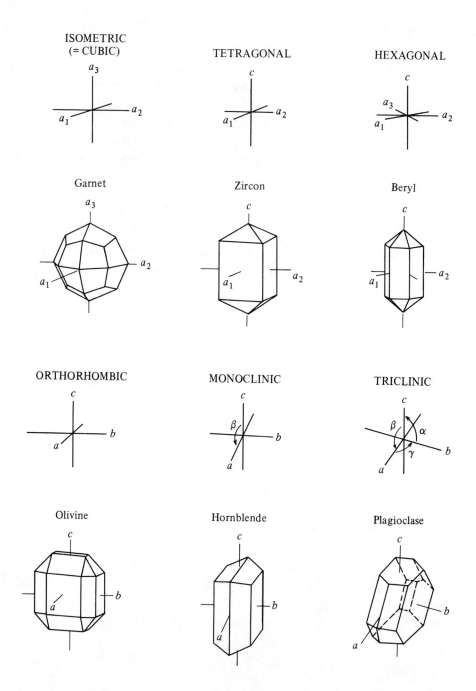

Figure 2.4. Crystal systems and representative crystals. All crystals may be described in terms of the six different sets of axes shown in this diagram. Where two or more axes are labled a, they have the same length – for example, in the hexagonal system, $a_1 = a_2 = a_3$, but the c axis has a different length. All angles are 90° (or 120° in the hexagonal system) except for those that are labeled with a Greek letter. An example of a relatively common mineral that crystallizes in each system is given beneath the appropriate set of axes.

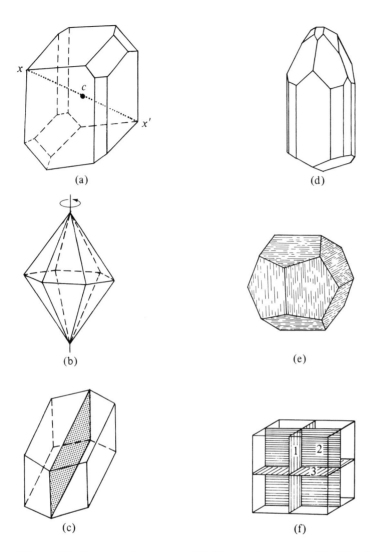

Figure 2.5. Elements of symmetry. (a) Center of symmetry. (b) Axis of sixfold symmetry. (c) Plane of symmetry. (d) Tourmaline crystal. There is no center of symmetry, but there are three planes of symmetry. (e) Pyrite crystal, exhibiting the commonly present striations. It has a center, several axes, and three planes of symmetry. (f) Schematic diagram showing the three planes of symmetry referred to in (e).

crystal presents the same appearance to a viewer following rotation through one of the following angles: 60°, 90°, 120°, or 180°. For example, if a crystal is rotated about a 180° axis of symmetry, the crystal will appear, to the stationary viewer, identical at the initial position (0°) and at 180°; such an axis is a twofold axis of symmetry. Similarly, if the appearances are the same after 120° rotation, the axis is a threefold axis of symmetry; if 90°, a fourfold axis; if 60°, a sixfold axis. In a cube, there are three fourfold axes of symmetry, each of which is perpendicular to, and through, the centers of the two opposite faces. A third element of symmetry is called a *plane of symmetry*. If a crystal can be divided into two exact mirror halves by a plane, that plane is called a plane of symmetry. For example, in a cube, a plane of symmetry passes through each pair of opposite edges. A given crystal can have only one center of symmetry or none, but it may have several axes and/or planes of symmetry or none.

In 1830 the German crystallographer Johann Friedrich Christian Hessel showed that all known crystal forms could be described in terms of thirty-two different groupings of symmetry elements. The groupings can be related to the six crystal systems in the following manner: In the cubic system, there are five symmetry classes; in the tetragonal system, seven classes; in the hexagonal system, twelve classes; in the orthorhombic system, three classes; in the monoclinic system, three classes; in the triclinic system, two classes.

The discovery of a mathematical system based on symmetry operations, rather than on axes and the familiar rules of geometry, intrigued crystallographers and mathematicians alike. Using symmetry operators, they could, in theory, start with a plane surface and then, by symmetry, generate every known crystal form.

The next question was obvious: If crystals are made up of tiny particles arranged in regular, three-dimensional arrays, could symmetry operations be used to describe all of the ways that space could be filled, according to each array? At first glance, the answer seemed to be yes. But suppose, as Huygens had suggested two hundred years earlier, that the tiny particles at the nodes of an array were not spheres, but instead were shaped like, say, a football. Obviously, if some of the footballs were to point in one direction and others in a second direction, the external crystal symmetry could still be described in terms of the thirty-two crystal classes, but additional information would be required to describe the orientations of the footballs. Considerations such as this ultimately led to the derivation of the 230 space groups that are now known to be needed to describe all of the ways that atoms can fill space. Remarkably, space group systematics were derived independently between 1885 and 1894 by three men: the Russian crystallographer Evgraf Stepanovich Federov, the German mathematician Arthur Moritz Schoenflies, and the amateur mathematician and semiretired English businessman William Barlow.

So the thirty-two crystal classes concern the symmetry of the external shapes of crystals, and the 230 space groups concern the symmetry of the internal arrays that describe the packings of atoms within crystals. The two are, of course, closely related.

Predictions arising from the six crystal systems and the thirty-two symmetry classes could be directly related to, and tested by work on, both natural and synthetic crystals. On the other hand, ideas about the 230 space groups were theoretical and could not be tested because there was then no way to "see" inside a crystal and thus to look at the arrays and orientations of internal particles. Thus, the work of crystallographers seemed to have come to a dead end at the close of the 19th Century.

Suddenly, however, as has happened so often in science, a new and entirely unsuspected avenue of research was opened. The new approach came from the discovery of X-rays, a discovery that turned out to be one of the greatest of all scientific discoveries. X-rays provided a way to examine and measure the internal atomic packings of crystals, and consequently those studies led to many of today's technological wonders, including such things as transistors and incredibly strong alloys. In addition, and perhaps of even greater significance, much of modern biophysics has developed from the discovery of X-rays. In fact, the very methods that were first used to determine the crystal structures of minerals are the methods now used to unravel the atomic structures of the chemical compounds in living bodies.

X-RAYS AND CRYSTAL STRUCTURES

X-rays are electromagnetic radiation, just as are light and radio waves. The difference is that the wavelength of X-rays is extremely short – so short, in fact, that they can penetrate solid bodies.

The German physicist Wilhelm Conrad Röntgen discovered X-rays in 1895. While carrying out experiments involving electrical discharges within partially evacuated glass vessels, "vacuum tubes," Röntgen noticed that an unknown radiation seemed to be passing through the black paper that surrounded the apparatus; he saw a nearby phosphorescent screen glowing brightly every time he ran his experiment. He did not know the nature of the strange rays that made the screen glow, so he called them X-rays, and that name is still applied to them, even though we now understand what they are. In 1901, Röntgen received the Nobel Prize in physics for his discovery.

Von Laue and the discovery of x-ray diffraction

The debate about X-rays was intense. Were they rays, like light, or were they streams of tiny particles?

In Röntgen's day, the diffraction properties of light were being studied intensively by means of diffraction gratings – very fine, very close parallel lines ruled on glass. This was done because diffraction occurs when a train of waves meets a series of fixed objects separated by a distance close to the wavelength. Light passing through an appropriate grating is diffracted and appears as a series of light and dark bands. From the width of the bands and the distance between the lines of a diffraction grating, one can calculate wavelengths. It was reasoned, therefore, that if X-rays were rays in the same sense that light is, X-rays also should be capable of being diffracted, and then it would be possible to determine their wavelengths. Try as they would, however, physicists could not make a grating fine enough to diffract X-rays.

Max von Laue, a young physicist working in Munich who was familiar with the work of crystallographers, reasoned that the geometric stacking of particles in crystals might serve as an exceedingly narrow grating that would be fine enough to diffract X-rays. However, a grating ruled on glass has two dimensions and gives diffraction bands, whereas a crystal is a three-dimensional grating and thus should produce diffraction spots. Two of von Laue's student assistants, Walter Friedrich and Paul Knipping, carried out an experiment devised by von Laue. They directed a beam of X-rays onto a crystal of copper sulfate dihydrate ($CuSO_4 \cdot 2H_2O$) and put a photographic film behind the crystal. When the film was developed, a series of diffraction spots was present. Von Laue was right: Crystals do diffract X-rays (Fig. 2.6). The year was 1912, and that discovery marked the birth of modern mineralogy.

Crystal structure

Von Laue's discovery aroused intense interest. Two British scientists, Sir William Henry Bragg and his son William Lawrence Bragg, immediately used x-ray diffraction to work out the stacking of atoms in two minerals: halite (common salt, NaCl) and sphalerite (the zinc mineral ZnS) (Plate 9).

The Braggs noted that von Laue's diffraction spots yielded two kinds of information: First, they could measure the positions of the diffraction spots, and from those measurements they could calculate the shape of the internal array of atoms by using the information known about space groups. Second, they could measure the intensities of the spots; some were strong, others less so, and the differences depended on the fact that some atoms were much more effective scatterers of X-rays than were others. Therefore, by combining the two kinds of information, they could determine both the geometry of the internal array of atoms and the sites where the different constituent atoms were located within the array.

The Braggs completed their epochal crystal studies in the same year as von Laue's discovery. Von Laue was awarded the Nobel Prize in physics in 1914, and the Braggs were jointly awarded the Nobel Prize in physics in 1915.

Crystal realms

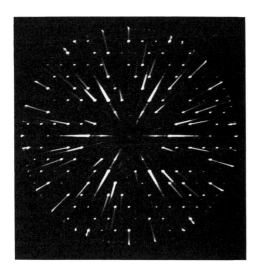

Figure 2.6. Diffraction of a beam of X-rays propagated along the c crystallographic axis of a crystal of apatite. The fact that the diffraction is recorded on a film as a series of discrete spots is proof that the constituent atoms of apatite are arranged in a specific geometric order. The distribution of the diffraction spots shows the symmetry elements – in this case, a sixfold symmetry around the c axis – and the spacings between atoms. (Photograph courtesy of D. R. Pecor)

From the time when the first crystal structures were worked out by the Braggs to the present, *x*-ray diffraction has been the means by which the atomic stackings inside crystals have been determined. Today, with the aid of computers, even extremely complicated structures can be deciphered relatively quickly.

THE PHYSICAL PROPERTIES OF CRYSTALLINE SOLIDS

While crystallographers and mathematicians of the 1800s concerned themselves with the shapes and geometries of crystals, mineralogists of the time were busy measuring other properties of minerals. The kinds of properties studied were the way minerals break, their densities, how they respond within a magnetic field, whether or not they fluoresce in ultraviolet light, their relative hardnesses (i.e., their resistance to scratching), and what happens to a ray of light when it passes through a homogeneous grain of a mineral.

Many of the properties turned out to be rather complicated, and as a result a whole new field of research, *crystal physics,* came into being. Most important, measurements of properties made on well-characterized minerals proved that minerals can be identified unambiguously by measuring one or more of their physical properties. That is to say, it is not necessary

Table 2.1. *The Mohs hardness scale*[a]

Hardness	Substance
10	Diamond
9	Corundum (e.g., ruby & sapphire)
8	Topaz
7	Quartz (e.g., amethyst & "rock crystal")
6	Orthoclase (a feldspar)
5	Apatite
4	Fluorite
3	Calcite
2	Gypsum
1	Talc

[a] On this scale, a jackknife and window glass have hardnesses of about 5½; a copper coin has a hardness of about 3½; one's fingernail has a hardness of about 2½.

to analyze a mineral grain chemically and determine its crystal structure by means of x-ray diffraction in order to give an unambiguous identification; other properties of a mineral are also specific and just as characteristic as the mineral's composition and its crystal structure.

The physical characteristics that are most useful and most widely employed for rapid identification of minerals are hardness, density, pattern of fracture, and the optical properties that can be determined either by passing light through a grain or by reflecting light from a polished surface of the mineral.

Hardness

A relatively simple but reliable way to distinguish between a diamond and a glass imitation is to try to scratch a piece of window glass with the unknown gem. A diamond is harder than window glass and will make a clear scratch; a glass imitation is a little softer than window glass and therefore will make no impression.

It is possible to rank order minerals in a sequence of relative hardness based on their ability to scratch one another. Diamond will scratch all other minerals and thus is the hardest. Corundum, which is better known by its gem varieties, ruby and sapphire, is the second hardest mineral. Corundum cannot scratch diamond, but it can scratch all other minerals. The first person to rank the common minerals on the basis of relative hardness was the previously mentioned German crystallographer, Friedrich Mohs (1773–1839), and even today mineralogists employ the Mohs hardness scale (Table 2.1) in one of the easily conducted mineral tests.

GEMSTONES

Gems were the first minerals to arouse human aesthetic senses. We know this because stones chosen and in some cases even shaped for adornment have been found in the most ancient graves.

Today, gemstones in museums, collections of crown jewels, and other exhibits attract the attention of millions of people every year. The mining, fashioning, and sale of gemstones constitute a multimillion-dollar industry.

Minerals and rocks that have been polished, or cut and polished, for use in jewelry are called gemstones (Plate 10). Most gemstones are used for personal adornment — for example, as necklaces, earrings, nose rings, finger rings, bracelets, brooches, or navel buds — but some are used to embellish decorative objets d'art. The famous "Blumenstrauss aus Edelsteinen" (floral bouquet of gemstones) presented in 1764 by Maria Theresa, Archduchess of Austria and Queen of Hungary and Bohemia, to her spouse, the Holy Roman Emperor Franz I, is an outstanding example of the latter use (Plate 11).

Many of the largest and most famous gemstones are displayed in museums. The Hope Diamond, now in the U.S. National Museum (Smithsonian Institution), and the crown jewels of England, in the Tower of London, are examples. But some of the large "gemstones" in museum exhibits, however magnificent, are really aberrant pieces fashioned primarily for the owner's bragging rights. Consider, for example, one of the world's heaviest faceted gemstones, the 22,892.5-carat (~10-lb) topaz in the Smithsonian Institution (Fig. 2.8): Granted, this fine, light golden yellow topaz is a beautiful showpiece, but certainly a cut and polished gem of this size,

Figure 2.7. Ancient seals. Minerals and rocks, such as chalcedony (upper right), lapis lazuli (bottom left and center), and soapstone (large, round specimen), were carved and used to identify the person who sealed certain ancient documents and communications. Later, these seals and similarly carved stones were also used for adornment. These early Egyptian and Greek seals are in the Metropolitan Museum of Art in New York City. (Photograph courtesy of American Federation of Mineralogical Societies)

Figure 2.8. This 22,892.5-carat (~10-pound) golden yellow topaz (NMNH #G9875) was presented to the U.S. National Museum of Natural History by the American Federation of Mineralogical Societies and Drs. Marie and Ed Borgatta of Seattle, Washington, in 1988. It was fashioned from a 26-pound, water-worn Brazilian crystal. (Photograph by D. Penland, courtesy of Smithsonian Institution)

about the size of an automobile headlight, could be used as a piece of jewelry only by a circus elephant or perhaps by a Clydesdale.

In order for a mineral or rock to be used as a gemstone, it must possess eye appeal, workability, durability, and preferably also rarity. If the stone has associated lore, so much the better. Indeed, the interest in lore is so great that the legends attached to some gemstones have been fabricated just to enhance merchandizing efforts.

More than one hundred minerals and rocks have been found in varieties suitable for use as gemstones. Only about two dozen, however, have found widespread use. The best known are diamond, corundum (ruby and sapphire), and beryl (emerald). Others that are well known are, alphabetically by mineral name, aragonite (pearl), beryl (aquamarine and morganite), chrysoberyl (alexandrite and "cat's-eye"), feldspar (especially "moonstone," "sunstone," and "amazonstone"), garnet, jadeite (jade), lapis lazuli, nephrite (jade), olivine (peridot), quartz (especially amethyst, citrine, agate, and "tiger's-eye"), spinel, spodumene (hiddenite and kunzite), topaz, tourmaline, turquoise, zircon, and zoisite (tanzanite). All of these are minerals, except lapis lazuli, which is a rock; lapis consists largely of the minerals lazurite (blue) and calcite (white), along with lesser amounts of pyrite and/or native gold. Two other natural substances that are often used as gemstones are amber (an amorphous fossil resin) and opal (a mineral gel); these are described briefly in Chapter 4.

Gemstones are sometimes labeled "precious" or "semiprecious," with diamonds, rubies, sapphires, and emeralds usually referred to as precious, and the other gemstones as semiprecious. This bipartite distinction has little meaning and is not really even acceptable; among other things, several of the so-called semiprecious gems are at least as expensive, weight for weight, as some of the so-called precious stones. In any case, one should never overlook the fact that the value of a gemstone is not inherent; rather, the value, whatever it is at any given time, is based on demand in the marketplace, which reflects such ephemeral things as fashion and national preferences.

Most transparent gemstones are cut and faceted. Facets are arranged so that light passing into the gem is reflected and refracted in such a way that the stone exhibits its best color and/or its greatest sparkle (usually referred to as its "fire"). Because color is so important, some gemstones are treated by heating or irradiation to improve their color and/or clarity.

Nearly all translucent and opaque gemstones are merely polished by tumbling or are cut and polished into dome-shaped cabochons, cameos, intaglios, or beads of diverse shapes. The choice is usually made in order to exhibit or emphasize one or more of the gemstone's appealing characteristics; for example, star stones, such as star sapphires, cat's-eyes, and tiger's-eyes, exhibit those effects best, if not exclusively, in cabochons, whereas turquoise is attractive no matter what its shape.

Some of the gemstones that are marketed today are fashioned from synthesized minerals or mineral-like substances, rather than naturally occurring minerals. These **synthetic gems** tend to be much less expensive than the natural gems they mimic. Unfortunately, some natural minerals can be distinguished from their synthetic analogues only by skilled experts employing fairly sophisticated instruments. This is so because both the natural and synthetic stones are the same substance, and thus ordinary chemical and physical tests cannot distinguish between them. Indeed, subtle and often difficult-to-detect internal hints of how and where a gem was formed must be employed, and only experts know how to make such distinctions.

A **simulated gem** differs from the substance indicated by its name – that is, a simulated ruby is neither a natural nor a synthetic red corundum. Instead, simulants are other natural minerals, other synthetic minerals, or even man-made glass. For example, a simulated ruby can be a natural red tourmaline, a synthetic red spinel, or red-colored glass. Glass simulants are often referred to as **paste**.

A reputable dealer will label gems by their correct names and honestly represent simulated versus synthetic versus natural gemstones. Unfortunately, some dealers do not. Therefore, when buying gemstones, three rules should be kept in mind:

1. Caveat emptor: Let the buyer beware, especially of "bargains."
2. Buy only from reputable dealers.
3. Remember that a gemstone name adorned with a modifying geographic adjective – for example, "Ceylonese peridot" – is very likely a simulant.

Density and specific gravity

Density is an important property of matter that is defined as the mass per unit volume. A mineral such as diamond, in which the atoms are tightly packed, has a relatively high density – for example, diamond is about 1.675 times as dense as graphite (see Fig. 1.1). Density, however, is not easily measured; a related property, specific gravity, is measured instead. *Specific gravity* is the ratio of the weight of a mineral to the weight of an equal volume of a liquid or gas. It turns out that this ratio is especially useful if water is chosen as the liquid standard, because the density of pure water is, by definition, equal to 1. Therefore, the specific gravity of a mineral when measured against water is numerically equal to its density. Because specific gravities of most minerals range from about 1.5 to as high as 19.5 and can be measured readily, this property was widely employed by mineralogists during the last century. Indeed, it is still employed today, especially by jewelers, who do not want to damage or destroy the gems they are checking. For the most part, however, specific gravity is now used much less widely than hardness or the two properties described next.

Mineral fracture

Huygens noted that crystals of Iceland spar tend to fracture along three planar directions when they are shattered. The three directions are not perpendicular; so the breakage fragments tend to be rhombohedral parallelepipeds. As mentioned earlier, Huygens reasoned that the planar directions of breakage must represent directions of weakness,

Gems, granites, and gravels

(a) (b)

*Figure 2.9. Mineral cleavage. Some minerals have no cleavage; others have up to six good cleavages. (**a**) Mica has one good cleavage that yields thin sheets. (**b**) Halite (salt) has three good cleavages at right angles to each other. (Photographs by B. J. Skinner)*

and hence directions of relatively easy breakage between the tiny particles that make up a crystal. His speculation was quite correct. Today, such preferred directions of planar breakage in a mineral are called *cleavage directions* (Fig. 2.9).

Cleavage directions in minerals were studied intensively during the 19th Century. This was prompted, at least in part, by the fact that all surfaces of the mineral grains in freshly broken rock specimens are breakage surfaces. Thus, it was recognized that a knowledge of how minerals break could be a highly useful aid in the identification of minerals within rock specimens. Today, a study of how a mineral breaks continues to be one of the first and quickest tests made when an unknown mineral is being studied.

Members of one well-known mineral group, the micas, have a single perfect cleavage. Members of the most common group of minerals on the face of the Earth, the feldspars, have two cleavages. Halite, which is the mineral name for common salt, has three cleavages; several minerals have four cleavages, and a few have as many as six. It is possible to identify many common minerals merely by determining the number of cleavages present and the angles between them.

A few minerals have no easy directions of breakage in their structures and hence have no cleavage directions. Examples of minerals that lack cleavage are quartz and garnet. Chiefly because of their lack of cleavage, both quartz and garnet are used widely as abrasives – for example, as grinding powders and for sandpaper. Even though they are not as hard as diamond or corundum, both of these minerals are tough because they do not cleave and do not readily break into smaller and smaller fragments.

Optical properties

Two common effects associated with light are *reflection*, which makes a smooth surface look shiny, and *refraction*, which makes a stick appear bent when it is dipped into water.

Reflection is not a very useful property for mineral identification, even though it can be employed for the study of opaque minerals. The field of study that uses reflected light is called reflected light microscopy; it is most often used for studies of ore minerals because so many of them are opaque.

Refraction, which is the property of light that makes it possible to make lenses for eyeglasses, happens because the speed of light changes whenever it passes from one medium, such as air, into another, such as water, glass, or a crystal. Refraction is a property widely used in the study of transparent minerals. Fortunately, more than 90 percent of all minerals are transparent.

Other optical effects arise in transparent crystals in which the packings of atoms are different in different directions, and consequently the velocities of light are different in different directions. As a result, the properties of light passing through such crystals differ with the direction of transmission, and by detecting those differences it is possible to get a measure of the atomic packings. On the other hand, the velocity of light is the same in all directions in glasses, liquids, gases, and crystals of the cubic system (e.g., halite).

The first mineral that attracted special attention because of its optical properties was Iceland spar. In 1669, the same year that Steno wrote his great treatise, another Danish scientist, Erasmus Bartholinus, commented on the fact that when a mark on a piece of paper was viewed through a crystal of Iceland spar, the mark appeared to be doubled (Fig. 2.10). As just mentioned, the phenomenon of the apparent bending of a stick in water is due to refraction. A similar phenomenon causes the doubling of a mark viewed through Iceland spar. In this latter case, the light ray is split into two rays, and the two rays are refracted differently; the doubling is termed *double refraction*.

It was soon discovered that crystals belonging to the cubic system exhibit only single refraction, whereas crystals of all the other (five) crystal systems exhibit double refraction. And, as one might suppose, glasses, as well as liquids and gases, also exhibit only single refraction.

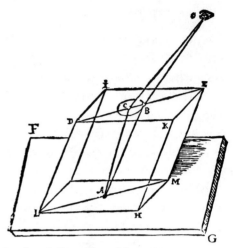

Figure 2.10. Figure from Experimenta Crystalli Islandici, *by Erasmus Bartholinus, published in Hafniae (Copenhagen, Denmark) in 1669. As can be seen, this diagram illustrates the movable image seen when one views something through a cleavage rhombohedron of Iceland spar (calcite).*

In the year 1828 the English scientist William Nicol carried out some important investigations that eventually led to an understanding of the split-image phenomenon in Iceland spar. He discovered that the effect occurs because a ray of light passing through a crystal exhibiting double refraction is divided into two parts that vibrate perpendicular to each other. Although a shaft of light that vibrates in only one plane appears to the eye as an ordinary ray of light, tests can readily be made to demonstrate the true direction of vibration. An easy test is to pass the ray of light emerging from a crystal through a sheet of Polaroid. Light passing through Polaroid also vibrates in only one direction. Indeed, all such rays are said to be polarized. When the true vibration directions in the mineral and the Polaroid are parallel, a maximum amount of light is transmitted; when the directions are perpendicular, no light is transmitted.

Several additional important discoveries were needed, however, before polarized light could be used to help identify minerals. Three steps were paramount: The first step came about as the result of a suggestion made in the early 1800s by an English physician, Thomas Young. Young recalled a statement by the great student of sounds and violins, Ernst Florenz Friedrich Chladni, to the effect that sound travels faster along wood fibers than across them. To Young, that suggested that the two rays of light that are polarized in perpendicular directions within minerals might also travel at different speeds; if they did, the phenomenon of double refraction could be explained. Young was correct. The second discovery concerned how light polarized by one crystal might be transmitted through a second polarizing crystal. As just mentioned, when the directions of vibration in two polarizing crystals are perpendicular, no light is transmitted; when the directions are at some angle other than 90°, however, some light is transmitted, and it was soon discovered that the amount of light transmitted is proportional to the angle of rotation between the planes of polarization of the two crystals. The third great discovery was that most minerals are transparent in thin slices. That fact was discovered in the mid-1800s by the English geologist Henry Clifton Sorby, who found that when a slice of rock only 0.03 mm thick is glued to a thin glass slide, most minerals are transparent.

It was but a short additional step to view thin sections of rocks in polarized light, and then to have the light pass through a second "polarizer" after it had passed through the thin slice of rock (Plate 12). Subsequently it was found that certain optical properties are characteristic properties of individual minerals. Thus, as might be expected, the study of minerals with a polarizing microscope became, and even in this day of highly sophisticated instruments remains, one of the single most powerful techniques available to the scientists who study minerals and rocks made up of minerals.

FURTHER READING

Bentley, W. A., and Humphreys, W. J., 1931, *Snow Crystals*. McGraw-Hill, New York, 227p.
An extremely fine collection of photographs of snow crystals.

Berry, L. G., and Mason, B., 1983, *Mineralogy: Concepts, Descriptions, Determinations* (2nd edition by R. V. Dietrich). W. H. Freeman, San Francisco, 561p.
Chapter 2 of this introductory text, a good introduction to crystallography, elaborates on the concepts treated in this chapter.

Burke, J. G., 1966, *Origins of the Science of Crystals*. University of California Press, Berkeley, 198p.
A concise scholarly history of crystallography from its beginning up to the days of x-ray crystallography.

Glasser, L. S. D., 1977, *Crystallography and Its Application.* Van Nostrand Reinhold, New York, 224p.
An intermediate-level text covering all aspects of crystallography, particularly x-ray crystallography.

Hallet, J., 1984, How snow crystals grow. *American Scientist,* 72, 582–9.
A summary about laboratory experiments and field observations dealing with the formation of snowflakes and frost.

Hurlbut, C. S., Jr., and Switzer, G. S., 1979, *Gemology.* Wiley, New York, 253p.
An introduction to gems and gemology.

Krashes, L., 1984, *Harry Winston, The Ultimate Jeweler.* Gemological Institute of America, Santa Monica, California, 206p.
The illustrations of jewelry are beyond compare.

McKie, D., and McKie, C., 1974, *Crystalline Solids.* Wiley, New York, 628p.
A general survey covering essentially all aspects of crystallography and the crystalline state, including optical properties.

Martin, D. D., 1987, Gemstone durability: Design to display. *Gems & Gemology,* 23, 63–77.
The durability aspect – especially as it bears on how some three dozen gemstones should and should not be set, repaired, cleaned, and displayed – is described.

Phillips, F. C., 1971, *An Introduction to Crystallography* (4th edition). Oliver & Boyd, Edinburgh, 351p.
This is the last edition of an excellent, widely used introductory text.

Schneer, C. J., 1983, The Renaissance background to crystallography. *American Scientist,* 71, 254–63.
A well-illustrated summary of the framework for crystallography that was laid during the Renaissance.

Smith, J. V., 1982, *Geometrical and Structural Crystallography.* Wiley, New York, 450p.
A well-written, comprehensive treatment covering essentially all aspects of crystallography.

PERIODICALS

Gems & Gemology. Gemological Institute of America, 1660 Stewart St., Santa Monica, California 90404. This trimonthly periodical is a standard for those interested in gemstones.

The Mineralogical Record. 4531 Paseo Tubutama, Tucson, Arizona 85715. This bimonthly publication is directed at the high-level amateur and professional mineralogist.

3

Mineral chemistry

Mineral chemistry has had practical applications for millennia. The smelting of tin and copper from their minerals and the combination of those metals to make bronze date back at least to 3000 B.C. The widespread prehistoric ceramic industry was founded on even more ancient practices. Minerals and rocks were the sources for many of the ingredients in alchemists' recipes. The list could go on.

Mineral chemistry as an investigative science is of much more recent origin. It is usually considered to have had its true beginning, as a branch of chemistry, during the early years of the 19th Century. Its attaining the status of a science is generally agreed to have depended on and closely followed two major discoveries and the development of accurate methods for quantitative chemical analysis. The discoveries were J. L. Proust's proposal of the law of constant composition in 1799 and John Dalton's enunciation of atomic theory in 1805.

The *law of constant composition* holds that all samples of a given substance contain the same chemical elements and that those elements are always present in the same proportions. This law seems so straightforward and self-evident that it is difficult for us to realize today how powerful and influential the law was when it was first proposed. Consider, however, the fact that in 1799 many of the chemical elements had not yet been discovered. Dalton's *atomic theory* proposed that all ordinary substances consist of atoms – that is, of small particles of matter – and that the law of constant composition could be explained if those atoms combined in fixed ratios. Accurate analyses eventually proved both Proust's law and Dalton's theory to be correct.

The births of modern mineralogy and of modern chemistry were essentially coincident and closely related. The early union between them has been, and continues to be, highly productive. Among other things, many chemical elements have been discovered as a result of chemical analyses of minerals. Two examples follow:

Chromium was discovered in 1797 by the French chemist Louis Nicolas Vauquelin. He found it in a lustrous mineral of a deep orange-red color called crocoite (a lead chromate) (Plate 13). In order to direct attention to the fact that chromium compounds tend to be highly colored, Vauquelin based the element's name on the Greek word *chromos*, meaning color. Indeed, both the red color of most rubies and the green color of most emeralds depend on the presence of trace amounts of chromium.

Hafnium, a rare metal, was discovered in 1922 following a comment made by Niels Bohr, a Nobel laureate physicist, to György von Hevesy and Dirk Coster. Von Hevesy, a Hungarian radiochemist who later also became a Nobel laureate, and Coster, a Dutch *x*-ray spectrographer, were working in Bohr's laboratory in Copenhagen. Bohr speculated that the gap directly below the element zirconium in a tabulation of chemical elements based on their properties should be filled by an element having properties similar to those of zirconium. Considering the well-known fact that elements with similar properties frequently occur together in minerals, von Hevesy and Coster examined spectra of several zirconium-bearing minerals and discovered trace amounts of the "new" element in some of those minerals. They named the element after the Latin name for Copenhagen, *Hafnia*.

CHEMICAL ELEMENTS AND COMPOUNDS

Gold, silver, copper, and platinum, as well as diamond and graphite and about two dozen other minerals, are made up of atoms of only one chemical element. Such single-element minerals are usually referred to as native chemical elements. Chemical elements are, as first postulated by the English natural philosopher Robert Boyle in 1661, the individual kinds of matter represented by a particular kind of atom.

Halite, quartz, the feldspars, calcite, clay minerals, most gemstones, most ore minerals, and about 3,500 other minerals are chemical compounds. Each of these minerals contains two or more chemical elements combined in fixed ratios.

A full description of a mineral requires knowledge of the chemical elements present, the proportions of the elements, and also the geometry of the packing of the atoms of the elements within the mineral's structure. These data, in turn, serve as the basis for an appreciation of the properties and characteristics of minerals. For example, the facts that some minerals are magnetic, whereas others are not (Fig. 3.1), and that all specimens of some minerals have the same color, whereas specimens of other min-

Mineral chemistry

Figure 3.1. Some minerals are magnetic. Magnetite (Fe_3O_4) was named on the basis of its strong magnetism. This variety of magnetite, which is called lodestone, is a natural magnet – that is, it will pick up objects made of iron, just as manufactured magnets do. (Photograph by G. K. McCauley)

erals may be of just about any hue, can be readily explained in terms of the chemistry and atomic structure of minerals (Plate 14).

Chemical symbols

Symbols are used as a shorthand, but exact, way of referring to chemical elements. Each chemical element has a symbol. Precursors to the symbols in use today appear to have been used at least as early as the 2nd Century A.D. (Fig. 3.2). Additional symbols came into use during the ensuing centuries. It was not until 1811, however, that the Swedish chemist Jöns Jakob Berzelius proposed the system of symbols now in use throughout the world (see Appendix 1).

Each of today's symbols comprises one or two letters, in most cases either the first or the first and second letters of the element's name. For some elements, the letters are those of the current name – for example, O for oxygen and Ca for calcium. For others, they are taken from the name of the element in another language, usually Latin – for example, Au for the Latin name for gold (*aurum*). The symbols are used to represent atoms and ions in chemical formulas and equations and also to represent the elements themselves in the texts of scientific publications.

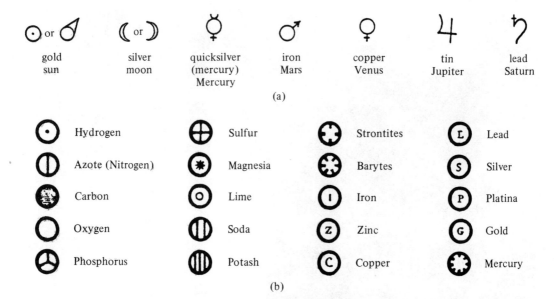

Figure 3.2. Early symbols used for chemical elements. **(a)** *Symbols used by ancients, including alchemists: Note that these symbols associated the known elements with the sun, the moon, and planets and that one of those early associations (Mercury with quicksilver) persists in present-day nomenclature.* **(b)** *Symbols used by, for example, John Dalton: Note that the symbols used for iron, zinc, copper, lead, silver, platinum, and gold were harbingers of the letters now used to designate the elements. The symbols used today are given in Appendix Figure A1.1.*

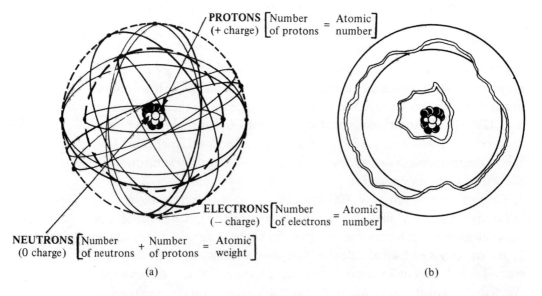

Figure 3.3. The atom. **(a)** *Conceptual model showing nucleus made up of protons and neutrons orbited by electrons.* **(b)** *Spherical shells indicating regions to which different electron orbits are restricted according to this "hard-shell" model.*

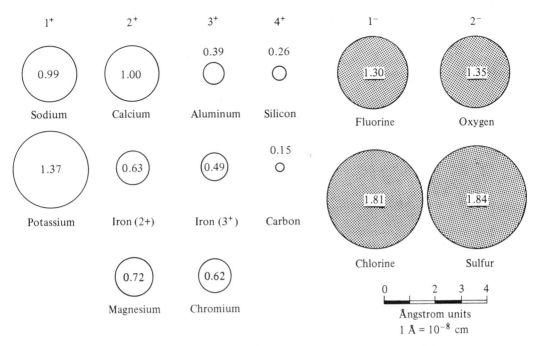

Figure 3.4. Different ions have different effective radii. This diagram illustrates the relative sizes of some of the common mineral-forming cations and anions. Numbers at top indicate valences.

Atoms

Atoms are the smallest particles that have all the properties of their respective elements. They are the fundamental units of each mineral's crystal structure. Although a picture of an atom as a hard sphere is not entirely realistic, it is nevertheless instructive for describing and illustrating the properties of both atoms and ions. We shall use it here (Fig. 3.3).

Each atom contains an extremely small nucleus that is made up of *protons,* each with a positive electrical charge, and, in most cases, also of electrically neutral *neutrons*. The nucleus is surrounded by one or more shells of negatively charged *electrons*. The whole acts as a sphere with an effective radius on the order of 1 angstrom (1 Å = 10^{-10} m = 0.000000003937 in. – that is, there are 2.54 billion Å in an inch or 10 billion Å in a meter). The actual radii of different atoms, however, differ; the effective radius of each is dependent on the nature of the atom, its state of ionization, and the manner in which it is linked to adjacent atoms and/or ions (Fig. 3.4).

A neutral atom has equal numbers of electrons and protons, and consequently the negative and positive charges are in balance. An *ion* is an electrically charged atom formed by the loss or gain of one or more electrons. There are two kinds of ions: *cations,* which have lost electrons and are positively charged, for example, Zn^{+2}, and *anions,* which have gained electrons and are negatively charged, for example, Cl^{-1}. There also are *cation groups*, for example, $(NH_4)^{+1}$, and *anion groups*, for example, $(CO_3)^{-2}$. Such groups, sometimes called *complex ions*, are made up of closely associated positively and negatively charged elements that are so tightly combined that they behave as units. The radius of an ion of an element differs from that of its neutral atom. This is true because any time a neutral atom releases an orbiting electron and becomes a cation, the positive electrical charge of the nucleus exceeds the negative charge of the remaining electrons, and as a consequence the remaining electrons are pulled in a little tighter by the nucleus so that the effective radius is decreased. With anions, the opposite effect occurs, and the radius is increased.

Valence

The term used to describe the number of lost or gained electrons – that is, the electrical charge of an ion – is *valence*. Two examples:

> The chlorine anion accepts an electron and has a negative valence of 1 (Cl^{-1}).
> The zinc cation gives up two electrons and has a positive valence of 2 (Zn^{+2}).

In the case of complex ions, the valences are added to give an overall valence; for example, the carbonate ion, which consists of C^{+4} and $3O^{-2}$ and is expressed as $(CO_3)^{-2}$, has a negative valence of 2.

Cations, anions, and complex ions within minerals are bonded so that their valences – that is, their positive and negative charges – are balanced. A compound in which all charges are balanced is said to be *stoichiometric*.

Bonding

The term *bonding* describes the linkages between adjacent atoms and/or ions in solids, liquids, and gases. Bonds are the forces that hold atoms in their places within the structures of minerals and other crystalline solids. In minerals, there are four main kinds of bonds: the metallic bond, the covalent bond (also called the homopolar bond), the ionic bond (also called the heteropolar or polar bond), and the van der Waals bond. The strengths of the different kinds of bonds range considerably.

Under the conditions of formation of a mineral, the structure that forms is almost always the one in which the constituent atoms and/or ions assume the arrangement and bonding that result in minimum internal energy. Thus, minerals may be said to have been formed in equilibrium with the external conditions of temperature, pressure, and chemical environment. Their charges are balanced. The packing of their constituent atoms and/or ions can be considered "rigid" – that is, so close that there is no "rattle."

METALLIC BOND. The metallic bond is predominant in the native metals and plays at least a minor role in many minerals that contain sulfur, selenium, tellurium, arsenic, antimony, or bismuth. Metallic bonding occurs when electrons can move readily from one ion to another – that is, there is an easy mobility of electrons. The mobility of electrons accounts for such diverse properties as the opacity and color of metals, their shiny luster, and their electrical and thermal conductivities.

COVALENT BOND. The covalent bond is predominant in only a very few minerals; diamond is the best-known example. In the diamond structure, a central carbon atom is surrounded by four other carbon atoms, each of which shares one electron with the central atom. Repetition of this pattern throughout the structure means that each diamond crystal is, in essence, one giant molecule in which each of the carbon atoms is covalent, that is, has a shared valence.

IONIC BOND. This bond is fundamental in about 90 percent of all minerals. In ionic compounds, each ion typically is surrounded by ions of opposite charge, with the number of ions dependent on their ionic radii and their charges. The structures of ionic compounds are therefore determined by both the sizes and the charges of their constituent ions; that is, ionic structures are controlled by the demands of both geometric stability and electrical stability. Geometric stability implies that both the relative ionic sizes and the modes of packing must result in the ions being held more or less rigidly within their structures. Electrical stability (neutrality) means that the positive and negative charges (the valences) must balance.

VAN DER WAALS BOND. Sometimes called the residual bond, the van der Waals bond is not easily explained, except in terms of quantum mechanics. Suffice it to say here that it is an attraction that results from electrical imbalances that depend on an off-centering of positively charged atomic nuclei within their negatively charged clouds of surrounding electrons. The attraction forces, which are weak, are thought to account for such things as the fact that graphite is a good lubricant.

The bond types are not mutually exclusive. In fact, the bonding in many crystalline substances involves more than one kind of bond. Take, for example, the bonds between the ions of silicon and oxygen that hold the complex silicate anion $(SiO_4)^{-4}$ together. This complex ion is present

Gems, granites, and gravels

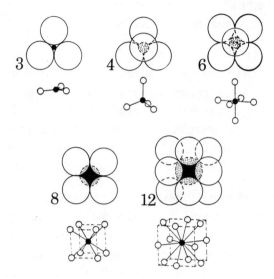

Figure 3.5. Coordination number of a cation defines the number and arrangement that anions (e.g., oxygen) may take around central cations of diverse sizes (see Table 3.1).

in all silicate minerals, the group that comprises the vast majority of all known minerals. The bonds in the $(SiO_4)^{-4}$ anion are neither purely ionic nor purely covalent; instead, they are partly ionic and partly covalent; that is, hybrid in nature.

An additional important fact with regard to bonding and minerals is that most common minerals are compounds of oxygen. Therefore, the fact that an oxygen anion has a large radius is significant. On a volume basis, oxygen exceeds all of the other elements in essentially all oxygen-containing minerals. This means that the structures of these minerals are determined largely by the arrangements of their oxygen anions and that the ions of the other elements, mostly smaller cations, merely fill interstices.

Ionic radii

Ionic radii are the values used to describe the sizes of ions. Within ionic structures, each cation tends to be surrounded by anions. As noted earlier, the number of anions grouped around each cation depends on their sizes as well as on their valences. The number of anions that fit around a given cation is generally referred to as the cation's *coordination number* (Fig. 3.5).

Because of the fact that oxygen is the most common anion in most minerals, the term *coordination number,* unless otherwise specified, refers

Table 3.1. *Relationships among coordination numbers, effective radii of cations acting as essentially rigid spheres, and their surrounding anions*

Coordination number	Arrangement of anions	Radius of cation
3	Corners of an equilateral triangle	0.15 – 0.22
4	Corners of a tetrahedron	0.22 – 0.41
6	Corners of an octahedron	0.41 – 0.73
8	Corners of a cube	0.73 – 1
12	Midpoints of cube edges	1

Note: See Figure 3.5.

to coordination with oxygen. If it is assumed that ions of a given element and valence act as spheres of a specific radius, the stable arrangement for a cation of any particular radius can be predicted (Table 3.1). This is so because the best fit is the one whereby adjacent anions just touch each other, are packed as efficiently as possible, and just touch the cation they surround. Five common coordination polyhedra are shown in Figure 3.5.

Some cations occur exclusively within particular coordination polyhedra. Others – those with radii near a theoretical size boundary between two kinds of coordination – may occur within either. Slight distortions in the geometry of coordination polyhedra are relatively common within minerals. Highly irregular and asymmetric polyhedra are rather rare.

MINERAL FORMULAS

Minerals, except the native elements, contain both cations and anions. In writing a chemical formula for them, the standard convention is that the constituent cations appear to the left, and the anions to the right. In addition, the positive charges of the cations must balance the negative charges of the anions. The examples given in Table 3.2 indicate how the scheme works.

Sometimes a mineralogist will speak of a mineral by name, but at other times in chemical terms. For example, halite is sodium chloride (NaCl), calcite is calcium carbonate ($CaCO_3$), and ilmenite is iron titanium oxide ($FeTiO_3$). If the mineral has H_2O in its formula, it is said to be a hydrate – for example, gypsum, with the formula $CaSO_4 \cdot 2H_2O$ could be termed a calcium sulfate dihydrate.

There is, however, a problem with the practice of referring to minerals

Table 3.2. *Examples of mineral formulas*

Formula	Cation(s)	Anion(s)	Mineral
NaCl	Na^{+1}	Cl^{-1}	Halite
$CaCO_3$	Ca^{+2}	CO_3^{-2}	Calcite
$FeTiO_3$	Fe^{+3}, Ti^{+3}	$3O^{-2}$	Ilmenite
MnO(OH)	Mn^{+3}	O^{-2} and $(OH)^{-1}$	Manganite
$KAlSi_3O_8$	K^{+1}, Al^{+3}, Si^{+4}	$2(SiO_4)^{-4}$	K-feldspar
$NaFe_3Al_6B_3Si_6O_{27}(OH)_4$	Na^{+1}, $3Fe^{+2}$, $6Al^{+3}$, and $3B^{+3}$	$3O^{-2}$, $6(SiO_4)^{-4}$, and $4(OH)^{-1}$	Schorl (a tourmaline)

simply in chemical terms: Some individual chemical formulas apply to more than one mineral species; that is, the same elements, in the same proportions, may have different crystal structures and thus constitute two or more different mineral species. One example of this phenomenon, which is termed *polymorphism*, has already been described and illustrated: the two forms of carbon (diamond and graphite). Another example is silicon dioxide (SiO_2) (Fig. 3.6), which is represented not only by the well-known mineral quartz but also by the minerals tridymite, cristobalite, coesite, and stishovite and, in addition, by the natural glass that has been given the name lechatelierite.

As can be seen in the tabulation of examples of formulas, most minerals contain two or more cations. Not shown are formulas in which two or more cations are enclosed within parentheses and separated by commas. This convention indicates that any combination of the elements within the parentheses may be present in a given atomic site within the crystal structure in a given specimen. This occurs because ions with similar charges and similar radii can substitute for each other. The phenomenon is called *ionic substitution*. A simple example is (Zn, Fe)S, a formula frequently given for the mineral sphalerite. The correct interpretation of this formula is that sphalerite, a zinc sulfide, may have part of its zinc substituted for by iron (Fe).

The tourmaline group of minerals, with the general formula $XY_3Z_6B_3Si_6O_{27}(OH)_4$, provides a more complex example. Each of the italicized letters – X, Y, and Z, which are not symbols for chemical elements – represents several possible elements that would be enclosed within parentheses in the complete chemical formula. Each of the italicized letters also represents a particular kind of site in the mineral structure – that is, a site that could be occupied by any one of the elements that would be enclosed within parentheses in the complete formula. The X sites may be occupied predominantly by Na^{+1} or Ca^{+2}, but may also include K^{+1}, or may be in part vacant; the Y sites may be occupied

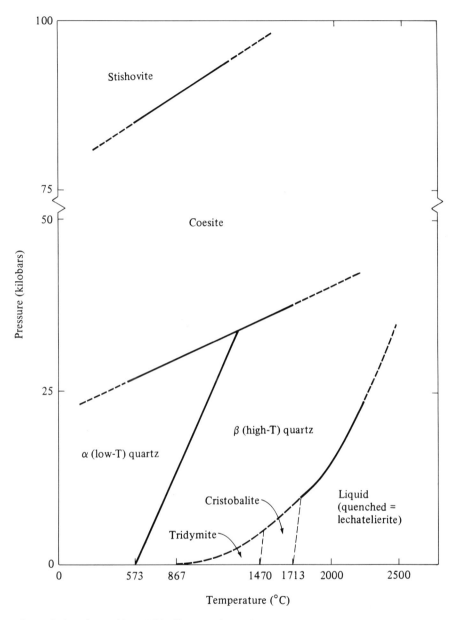

Figure 3.6. Polymorphism. This diagram shows the temperature–pressure conditions under which some of the diverse mineral species having the formula SiO_2 may form and be stable. Some of the polymorphic forms can persist in a metastable form under other, including atmospheric, conditions and thus serve as geothermometers and/or geobarometers – that is, they can indicate the general temperature–pressure conditions that existed when they were formed.

predominantly by Fe^{+2}, Mg^{+2}, Al^{+3}, Fe^{+3}, Li^{+1}, or Mn^{+2}; the Z sites may be occupied predominantly by Al^{+3}, Fe^{+3}, or Cr^{+3}, but may also include Mg^{+2} and V^{+3} in noteworthy percentages.

Most minerals contain small amounts of elements not indicated in their formulas. These elements, generally termed *trace elements*, usually are present at only a few parts per million (ppm). Many of these elements seldom, if ever, occur as major components in minerals. Some geochemists refer them as being dispersed, meaning that they usually substitute in small quantities for some of the major constituents in one or more of the relatively common minerals. As an example of trace-element distribution, all but 19 of the 91 naturally occurring elements have been recorded as occurring in tourmaline group minerals, and 52 of them only as trace elements. Because of this extraordinarily large number of trace elements, W. L. Bragg described tourmaline as "one of nature's catch-all or garbage-can minerals."

Mineral classification

Like the classification of other natural things, that of minerals involves the grouping of entities having similar characteristics and the separation of groups with different characteristics. These classifications provide order in arrays of data, thus making the mineralogical data easier to interpret and use.

The mineral classification now used by nearly all mineralogists is based on chemical compositions. As might be guessed from earlier statements, the major categories, except for the native elements, are based on the anion contents of the minerals (Table 3.3). This, of course, serves to emphasize the fact that minerals' structures, and consequently their characteristic properties, are largely dependent on the packing of their larger anions, whereas their smaller cations, of appropriate radii and valences, are only minor constituents, both volumewise and so far as controlling many of their properties. Consider, for example, that calcite ($CaCO_3$) and magnesite ($MgCO_3$) have rather similar properties, whereas calcite ($CaCO_3$) and fluorite (CaF_2) are very different.

Some of the major categories can, of course, be subdivided. For example, as shown in Table 3.3, the silicates usually are divided into six subclasses according to the structural linkage of their $(SiO_4)^{-4}$ complex anions. This sharing of two or more $(SiO_4)^{-4}$ anions to form even larger complex anions is called *polymerization* (Fig. 3.7).

As the classification indicates, the structures of diverse silicate minerals comprise cases where the $(SiO_4)^{-4}$ ions, commonly referred to as silica tetrahedra, join to form $(Si_2O_7)^{-6}$, $(Si_3O_9)^{-6}$, $(Si_4O_{12})^{-8}$, and $(Si_6O_{18})^{-12}$ groups.

The only rule governing polymerization of silica tetrahedra is that two adjacent tetrahedra can share only one oxygen; that is, tetrahedra can be joined only at their corners, never along their edges or faces. There is,

Table 3.3. *The major mineral categories*

Class	Example (formula)
Native elements	Gold (Au)
Sulfides (including selenides and tellurides)	Galena (PbS)
Sulfosalts	Tetrahedrite $[(Cu, Fe)_{12}Sb_4S_{13}]$
Oxides	Hematite (Fe_2O_3)
Hydroxides	Gibbsite $[Al(OH)_3]$
Halides	Fluorite (CaF_2)
Carbonates	Calcite $(CaCO_3)$
Nitrates	Nitratine $(NaNO_3)$
Iodates	Lautarite $[Ca(IO_3)_2]$
Borates	Borax $[Na_2B_4O_5(OH)_4 \cdot 8H_2O]$
Sulfates (including selenates, etc.)	Barite $(BaSO_4)$
Chromates	Crocoite $(PbCrO_4)$
Phosphates (including arsenates, etc.)	Turquoise $[CuAl_6(PO_4)_4(OH)_8 \cdot 5H_2O]$
Antimonates (including antimonites, etc.)	Swedenborgite $[NaBe_4(SbO_7)]$
Vanadium oxysalts	Carnotite $[K_2(UO_2)_2V_2O_8 \cdot 3H_2O]$
Molybdates and tungstates	Scheelite $[Ca(WO_4)]$
Organic compounds	Whewellite $(CaC_2O_4 \cdot H_2O)$
Silicates	
Nesosilicates	Forsterite (Mg_2SiO_4)
Sorosilicates	Hemimorphite $[Zn_4(Si_2O_7)(OH)_2 \cdot H_2O]$
Cyclosilicates	Beryl $[Be_3Al_2(Si_6O_{18})]$
Inosilicates	Diopside $[CaMg(SiO_3)_2]$
Phyllosilicates	Kaolinite $[Al_2(Si_2O_5)(OH)_4]$
Tectosilicates	Quartz (SiO_2)

however, no requirement that polymerized anions must be discrete, closed rings; that is, infinitely polymerized chains, sheets, and frameworks are possible. Consequently, there also are minerals that include such arrangements as chains of tetrahedra that give the anion $(SiO_3)_n^{-2}$, sheets of tetrahedra that give the anion $(Si_2O_5)_n^{-4}$, and three-dimensional frameworks that give $(SiO_2)^0$.

It is interesting that for most of the minerals in which large polymerized anions occur, each contains only one kind of silicate anion – for example, beryl contains only the $(Si_6O_{18})^{-12}$ six-member rings. A few silicate minerals, however, do contain two different kinds of silicate anions – for example, vesuvianite

$Ca_{10}Mg_2Al_4(SiO_4)_5(Si_2O_7)_2(OH)_4$

contains both $(Si_2O_7)^{-6}$ and $(SiO_4)^{-4}$ groups.

In some cases, even further subdivisions have been made by grouping minerals that are closely related chemically and/or structurally or even

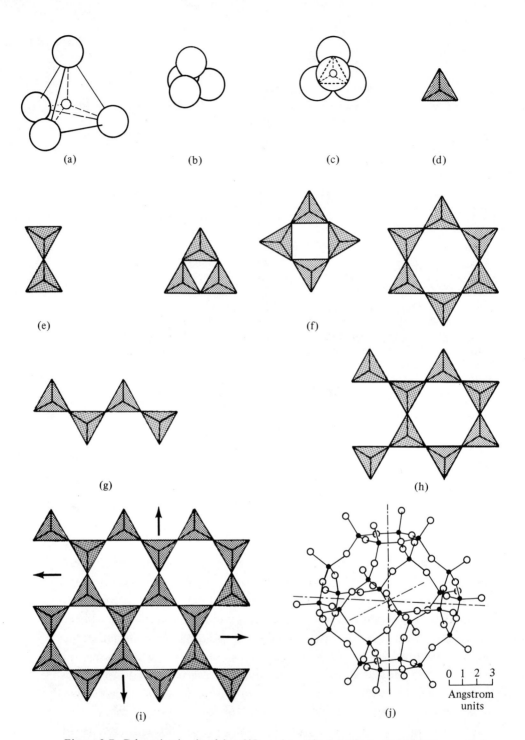

Figure 3.7. Polymerization involving SiO_4 anions. Sharing of two or more SiO_4 tetrahedra leads to even larger, more complex anions. **(a)** *Silica tetrahedron, expanded view, showing large oxygen ions at corners and small silicon ion in center.* **(b)** *Tetrahedron as it really is.* **(c,d)** *Diagrams showing derivation of the schematic tetrahedron used in other diagrams.* **(e,f)** *Isolated groups.* **(g)** *Single chain.* **(h)** *Double chain.* **(i)** *Triple chain.* **(j)** *Three-dimensional network with solid dots for silicon and open circles for oxygen. These groupings are fundamental to silicate minerals that are referred to as* **(d)** *neosilicates,* **(e)** *sorosilicates,* **(f)** *cyclosilicates,* **(g,h)** *inosilicates,* **(i)** *phyllosilicates, and* **(j)** *tektosilicates.*

by grouping them on the basis of some particular origin. Species so related usually are considered to constitute mineral groups – for example, the garnets, the micas, the amphiboles, the pyroxenes, and the zeolites.

A few mineral species include varieties that are so common that they have been given names. Color is the most common basis for naming diverse varieties. Quartz (SiO_2), for example, includes the well-known color varieties amethyst, rose quartz, smoky quartz (cairngorm), citrine, and rock crystal, as well as several kinds of very fine grained quartz that are grouped under the general name chalcedony; chalcedony, in turn, includes agate, bloodstone, carnelian, chrysoprase, and sard (Plate 15). In fact, many gemstones are simply color varieties of relatively common minerals. For example, ruby and sapphire are red and blue varieties, respectively, of the relatively common mineral species corundum (Al_2O_3), and emerald and aquamarine are green and blue varieties of beryl ($Be_3Al_2Si_6O_{18}$). The colors of these gemstones (and several others) are due to trace elements that are not essential constituents of the mineral species.

THE NAMING OF MINERALS

The names of some minerals are so ancient that their origins are lost in the dim mists of antiquity. In the 1st Century A.D., Pliny listed a number of the native elements, together with a few common ore minerals and gem minerals. Some of the names he recorded – for example, pyrite and gypsum – have persisted.

In the latter part of the 18th Century, a number of rival systems of nomenclature were proposed. For example, Carolus Linnaeus applied the same binomial nomenclature that he developed for plants and animals. Thus, his genus *Baralus* included *B. ponderosus, B. prismaticus, B. fusilis,* and *B. rubefaciens*. Today, we call these minerals barite ($BaSO_4$), celestite ($SrSO_4$), witherite ($BaCO_3$), and strontianite ($SrCO_3$), respectively. By the middle of the 19th Century, however, nearly everyone dealing with minerals had adopted a system whereby each mineral had only a single name. This is the system we use today.

The names themselves are of diverse origins. Some have been derived from Greek or Latin words that afford certain information about the mineral – for example, a mineral's color (albite, from the Latin *albus,* white), its density (barite, from the Greek *barys,* heavy), or its crystal form (sphene, from the Greek *sphen,* a wedge). Other names have come from stems relating to the mineral's chemical composition – for example, zincite (ZnO), calcite ($CaCO_3$), and even babefphite, a name based on the mineral's chemical formula: $BaBe(PO_4)(F, O)$. Still other mineral names have descriptive suffixes – for example, the -clase (from the Greek *klasis,* fracture) in such names as orthoclase and plagioclase.

Unfortunately, descriptive terms can be used for only a few minerals. This is so because there are insufficient distinctive properties to which attention can be directed merely by a simple name. Consequently, throughout the 20th Century, newly discovered minerals have tended to be named for such things as the localities where they were first found, or for people – generally scientists, but also mineral collectors, public figures, and others. Examples of these kinds of names are

> anglesite for the island of Anglesey off the coast of Wales,
> aragonite for Aragon, Spain,
> zunyite for the Zuni mine, San Juan County, Colorado,
> cliffordite for Clifford Frondel,
> fleischerite for Michael Fleischer,
> tombarthite for Tom F. W. Barth, and
> rooseveltite for Franklin Delano Roosevelt.

There is even a skinnerite, which is named after Brian J. Skinner, as well as a dietrichite, which is not named for R. V. Dietrich but for G. W. Dietrich, the 19th-Century Bohemian chemist who first analyzed the mineral. (In any case, these two names seem appropriate – skinnerite is an ore mineral, and dietrichite is a vitriolic efflorescence.)

Two or more names have been applied to a few individual species. When such cases are recognized, the standard procedure is to adopt the name that was applied first; this is known as the rule of priority. Other difficulties relating to mineral nomenclature also arise from time to time. The International Mineralogical Association's Subcommission on New Minerals and Mineral Names is recognized as responsible for resolution of all questions relating to mineral names and nomenclature.

Number of minerals.

This is continually changing: New minerals are being discovered, and old minerals are sometimes discredited. Currently, approximately 3,500 minerals are accepted as valid species.

Additional species are being discovered at a rate of about eighty minerals per year. Most of the discoveries made in recent years have depended on the availability and use of relatively new technologies, particularly devices for analyzing microscopic mineral grains. One of these devices, the electron microprobe, can analyze grains only 10,000 Å (= 1 micron = 0.001 mm = 0.00003937 in.) across.

Mineral chemistry

HOW MANY MINERALS ARE THERE?

The simple truth is that we do not know. Indeed, we think it would be imprudent even to hazard a prediction.

Consider the history of the growing number of recognized mineral species:

During the 1830s, James Dwight Dana catalogued all the known minerals and came up with just a few less than 600 species.

Near the end of the century, in the 1890s, Edward Salisbury Dana repeated his father's task and recorded about 800 species.

During the early 1940s, a distinguished German mineralogist, Hugo Strunz, listed about 1,400 species.

By the mid-1980s, Michael Fleischer listed about 3,500 species.

And there seems to be no suggestion of any decrease in the rate of discovery, which has amounted to about eighty species per year in the 1980s.

When these kinds of data are considered, it becomes clear that the rate of discovery has increased over the years. The changes in the rate appear to be attributable to three main factors:

1. the widespread use of x-ray powder diffraction methods to identify and describe minerals,
2. the increase in the number of geologists and amateur collectors who have supplied more and more specimens for identification, and
3. the increased use of a device called the electron microprobe analyzer, which permits chemical analysis of extremely small mineral grains, many of which previously were overlooked or were too small for appropriate analyses.

As might be expected, disagreements can arise as to what is and what is not a new mineral species. Indeed, the International Mineralogical Association has a Commission on New Minerals and Mineral Names that considers the validity of newly discovered minerals and disagreements relating to mineral names. Fortunately, nearly all mineralogists accept the decisions of this commission; nonetheless, a few disagreements persist. But even considering the uncertainties that these disagreements introduce so far as exactly how many minerals had been discovered by any given date, it seems safe to say that the numbers of minerals given on such lists as Fleischer's have been well within ±5 percent of the correct number.

In any case, consider the new minerals that may be discovered as more and more of the unusual chemical regimes of our solar system and other parts of the universe are sampled in the future. It may well be that the total number of minerals will lie well beyond even the conjectures that sanguine optimists of the mineralogical profession might suggest today.

THE FORMATION OF MINERALS

Minerals form whenever and wherever the appropriate temperature, pressure, and chemical environment exist for a long enough period of time for the constituent atoms to become ordered into a crystalline structure. Some minerals form only within very limited ranges of temperature and

Gems, granites, and gravels

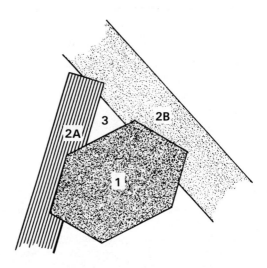

Figure 3.8. Perfection of crystal form may be described by the terms euhedral, subhedral, and anhedral. Grains like the one marked "1" are termed euhedral; *bounded by their own crystal faces, these grains appear to have been crystallized when nothing prevented them from developing all their own crystal faces. Grains such as the one marked "3" are termed* anhedral; *with none of their own crystal faces, grains such as these appear to have crystallized within the restricted spaces left after their surrounding minerals had already completed their growth. Grains like those marked "2A" and "2B" are termed* subhedral; *these grains have some, but not all, of their own crystal faces. (Of those shown, grain 2A appears to have had its growth restricted only by the euhedral grain 1, whereas 2B appears to have had its growth restricted by both the euhedral grain 1 and the apparently preexisitng subhedral grain 2A.) Some geologists use other terms to describe the same relationships – for example, idiomorphic or automorphic (for euhedral), hypidiomorphic or hypautomorphic (for subhedral), and allotriomorphic or xenomorphic (for anhedral).*

pressure; other minerals form over rather broad ranges of conditions. The former minerals are of special interest to the geologist because they may be useful as geothermometers and/or geobarometers – that is, they may indicate the temperature and/or pressure conditions that existed when and where they were formed.

Most minerals are formed as a result of one of the following processes:

1. crystallization from a magma (rock melt),
2. deposition from an aqueous solution,
3. condensation from a gaseous fluid, or
4. reaction in response to changes in temperature and pressure.

These same processes can be replicated in the laboratory to determine the conditions under which minerals may have originated and also to establish the conditions and processes that will yield desired synthetic minerals.

When the conditions of formation are such that "growth" is relatively slow and not physically constrained by grains of other minerals, well-developed crystals – sometimes referred to as *euhedral* crystals – often result. These crystals become the specimens that are prized by collectors and frequently are displayed in museums (Plate 16). On the other hand, where the growth of a crystalline grain is impeded, only a few, or none, of the crystal faces will develop. The resulting grain shapes are termed *subhedral* or *anhedral* (Fig. 3.8). Most rocks are made up largely of anhedral and subhedral grains.

MINERAL OCCURRENCES

Minerals, as already mentioned, are the predominant constituents of most rocks and loose aggregates, such as soils, dust, sand, and gravel. Minerals also occur in veins and in ore deposits, most of which are, in essence, only rather special kinds of rocks. These three kinds of geologic materials – rocks, loose aggregates (soils, dusts, and muds), and ores – are treated in Chapters 4, 5, and 6, respectively.

FURTHER READING

Berry, L. G., and Mason, B., 1983, *Mineralogy: Concepts, Descriptions, Determinations* (2nd edition by R. V. Dietrich). W. H. Freeman, San Francisco, 561p.
  Much of this introductory textbook relates directly to the information presented in this chapter.

Cotton, F. A., and Wilkinson, G., 1976, *Basic Inorganic Chemistry.* Wiley, New York, 579p.
  This is a good stepping-stone to the same authors' Advanced Inorganic Chemistry (4th edition) of 1980, also published by Wiley.

Fleischer, M., 1987, *Glossary of Mineral Species* (5th edition). Mineralogical Record, Tucson, 234p.
  All accepted minerals, as of the publication date, and their chemical compositions and crystal systems are listed.

Hampel, C. A., 1968, *The Encyclopedia of the Chemical Elements.* Reinhold, New York, 849p.
  This fine volume includes information on the discovery, occurrence, properties, and uses of each element.

McQuarrie, D. A., and Rock, P. A., 1987, *General Chemistry* (2nd edition). W. H. Freeman, San Francisco, 876p.
  This beautifully illustrated chemistry text is a fine introduction to such topics as atoms, ions, and chemical reactions.

Mitchell, R. S., 1979, *Mineral Names – What Do They Mean?* Van Nostrand Reinhold, New York, 229p.
  A collection of mineral names and their probable derivations.

Roberts, W. L., Rapp, G. R., Jr., and Campbell, T. J., 1990, *Encyclopedia of Minerals* (2nd edition). Van Nostrand Reinhold, New York, xxiii + 979p.
A color-illustrated rundown on properties and a few occurrences for many of the known minerals.

PERIODICALS

The American Mineralogist. This bimonthly publication is the professional journal of the Mineralogical Society of America, 1625 I Street, N.W., Suite 414, Washington, D.C. 20006.

The Canadian Mineralogist. This quarterly publication is the professional journal of the Mineralogical Association of Canada, Department of Mineralogy and Geology, Royal Ontario Museum, Toronto, Ontario M5S 2C6, Canada.

Mineralogical Magazine. This bimonthly publication is the professional journal of the Mineralogical Society, 41 Queen's Gate, London SW7 5HR, United Kingdom.

Gems, granites, and gravels

Plate 1. Diamond: puzzling mineral and prized gem. Diamond crystals are commonly shaped like this, the Oppenheimer diamond – NMNH #117538. (Photograph by D. Penland, courtesy of Smithsonian Institution)

Plate 2. This necklace, one of the highlights in the Smithsonian Institution, features the famous 67.9-carat Victoria Transvaal *diamond as its pendant. The stone from which it was cut was found in South Africa – NMNH #G7101. (Photograph by D. Penland, courtesy of Smithsonian Institution)*

Plate 3. Synthetic diamonds grown at ultrahigh pressures in the laboratory. Such diamonds, used as abrasives by industry, now account for more than two-thirds of the world's diamond consumption. Natural diamonds, however, are still preferred as gemstones. (Courtesy of General Electric Company, Specialty Materials Department)

Plate 4. Sands are loose aggregates of minerals and/or rocks. Although most sands consist largely of quartz, they also include several other minerals that are resistant to weathering and erosion. This group of sand grains – which includes such minerals as apatite [$\sim Ca_5(PO_4)_3Cl$] (white), cassiterite (SnO_2) (orange), corundum (Al_2O_3) and kyanite (Al_2SiO_5) (blue), pyrite (FeS_2) (golden brown), and the pyroxene diopside ($CaMgSi_2O_6$) (green) – was extracted from a sand by panning; the grains are shown here as seen in reflected light, through a microscope. [Photomicrograph from Guigues, J., and Devismes, P., 1969, La prospection minière à la batée dans le Massif Armorican. Mem. Bureau de Recherches Géologiques et Minières, Orléans–La Source (Loriet), France]

Plate 5. Example of a consolidated mineral assemblage. This distinctive rock from Gore Mountain, New York, is a metamorphic rock called garnet gneiss. Garnet crystals (red) are surrounded by collars of hornblende (black) set in a matrix of plagioclase feldspar and mica. (Photograph by Robert J. Tracy)

Plate 6. Colorful groups of crystals such as these are displayed in some of the finer museum collections. **(a)** Veszelyite [$(Cu, Zn)_3(PO_4)(OH)_3 \cdot 2H_2O$] – NMNH #148368. This group of crystals from Montana constitutes a "one in a million specimen" of the quality and rarity so highly sought by mineral collectors. **(b)** This specimen consists of pink rhodochrosite ($MnCO_3$) crystals on top of a mass coated with small quartz (SiO_2) crystals – NMNH #125901. (Photographs by D. Penland, courtesy of Smithsonian Institution)

Plate 7. Ice crystals on a window in southwestern Virginia. As can be seen by comparing this photograph with Plate 8, there is a remarkable similarity in the appearance – especially the luster and water-clear quality – of ice crystals and colorless quartz crystals. (Photograph by J. W. Murray)

Plate 8. Colorless quartz crystals from the Jeffrey Stone Quarry, Arkansas – NMNH #R12804. (Photograph by Chip Clark, courtesy of Smithsonian Institution)

Gems, granites, and gravels

(a)

(b)

*Plate 9. Crystal models of **(a)** halite (NaCl) and **(b)** sphalerite (ZnS), the first two mineral structures worked out by the Braggs. In the halite model, the red balls represent sodium (Na) and the white ones chlorine (Cl); in the sphalerite model, the gray balls represent zinc (Zn) and the yellow ones sulfur (S). Crystal models such as these are instructive in that they show the relative positions of the constituent atoms. (Photographs by David Darst; models assembled by Klinger Educational Products Corp., College Point, New York)*

Gems, granites, and gravels

Plate 10. This photograph was taken in 1949 to publicize the Court of Jewels. On her forehead, the model is wearing the famous Hope blue diamond, which is the central attraction of the gem display of the Smithsonian Institution; the Earl of Dudley necklace and the Inquisition necklace are around her neck; the Indore pear shapes lie on either side of the Star of the East; the Jonker diamond is at her waist; the 337-carat Catherine the Great sapphire graces her wrist; and the Mabel Boll and McLean diamonds are on her left hand. It is thought that if these gems were put on the market today, they would bring in excess of 250 million dollars. Some of them are virtually priceless, not only because of their size and quality but also because of the lore and history associated with them. (Photograph by permission, from Harry Winston: The Ultimate Jeweler; *© Harry Winston, Inc.)*

Plate 11. "Blumenstrauss aus Edelsteinen" (= flower bouquet of gems). This famous piece – which consists of 2,700 gems, including amethyst, chalcedony, chrysolite, diamond, emerald, garnet, lapis lazuli, ruby, sapphire, spinel, topaz, and turquoise – was presented by the Empress Maria Theresa to her husband, Franz I, in 1764. According to Dr. G. Niedemayr of the museum where the "bouquet" is housed, the designer was probably Lautensack of Frankfurt, but as now exhibited it was rearranged by the Viennese lapidary G. Gasterstädt in 1837. (Photograph courtesy of the Naturhistorisches Museum, Wein)

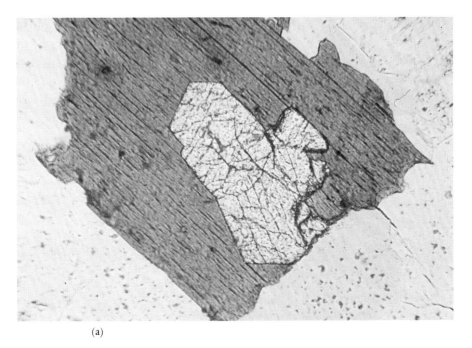

(a)

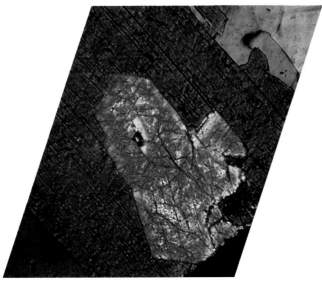

(b)

*Plate 12. Thin section of a specimen from the Mount Airy granodiorite from North Carolina. As noted in the caption of Figure 4.1, thin sections are slices of rock precisely cut and ground to have a thickness of 0.03 mm. When these slices are studied with petrographic microscopes, the minerals can be identified and the interrelationships among them may be seen. Relations such as these provide the kinds of evidence that permit petrologists to interpret the origins of rocks and their constituent minerals. **(a)** Thin section as viewed in ordinary light. **(b)** The same thin section as viewed between crossed polarizers. The mineral near the center is epidote; the brown mineral around it is biotite, a brown mica; and the surrounding light-colored minerals are feldspars. (Photographs by R. V. Dietrich)*

Plate 13. Crocoite ($PbCrO_4$). This is the mineral in which the element chromium (Cr) was discovered. This specimen is from the Dundas District, Tasmania, Australia. (Photograph by D. Penland, courtesy of Smithsonian Institution)

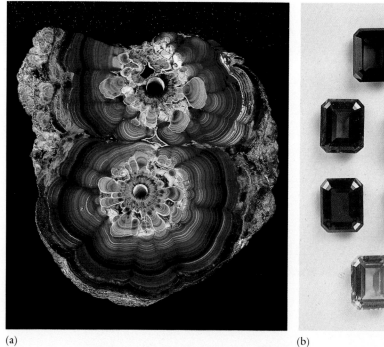
(a)

(b)

Plate 14. (a) Malachite $[Cu_2(CO_3)(OH_2)]$ and azurite $[Cu_3(CO_3)_2(OH)_2]$, which are green and blue, respectively, are examples of the many minerals each of which has its own constant (i.e., inherent) characteristic color. This specimen is from Copper Queen Mine, Bisbee, Arizona. (Photograph by S. C. Chamberlain of specimen in Pinch Mineral Collection of the National Museum of Natural Sciences, Ottawa, Canada; reproduced by courtesy of the museum) (b) Corundum is an example of the many minerals each of which is essentially colorless when pure but may exhibit several different (i.e., exotic) colors when they include, for example, certain trace elements. (Photograph by W. Sacco)

(a)

(b)

(c)

*Plate 15. Quartz is a mineral that has several named varieties: These three are often used in jewelry. (**a**) Smoky quartz, also called cairngorm – NMNH #114606. (Courtesy Smithsonian Institution) (**b**) Amethyst – NMNH #C6647. (Photograph by V. E. Krantz, courtesy Smithsonian Institution) (**c**) Agate, a variety of cryptocrystalline quartz, usually referred to as chalcedony. (Photograph by B. J. Skinner)*

Gems, granites, and gravels

Plate 16. Gold is quite different from "fool's gold." **(a)** Gold is malleable, has a high specific gravity (19.3), and a hardness of 2½–3. This fine specimen, from Gross Valley, California, is NMNH #121297. **(b)** Pyrite ("fool's gold") is brittle, has a much lower specific gravity (5.01), and a hardness of 6–6½. This spectacular group of cubic crystals, from Spain, is NMNH #R18657. (Photographs by D. Penland, courtesy of Smithsonian Institution)

Gems, granites, and gravels

Plate 17. Obsidian, a glassy igneous rock formed by the rapid cooling of lava. This specimen is from the Jemez Mountains, New Mexico. The reddish-colored spots are due to oxidation. (Photograph by W. Sacco)

Plate 18. Multicolored, gem-quality opal from Virgin Valley, Nevada. Note the fabulous colors and their arrangement. (See also Figure 4.3.) (Photograph by S. C. Chamberlain)

Plate 19. Flowing lava seen from the air at night during a 1984 eruption of Kilauea Volcano, Hawaii. Very fluid lava, such as that seen here, can flow down steep slopes faster than people can run! (Photograph courtesy of S. C. Porter)

Plate 20 (facing). Half Dome in Yosemite National Park is made up of an intrusive igneous rock of granitic composition. Thousands of feet of previously overlying rocks had to be removed by weathering and erosion to expose this mass. (Photograph by R. V. Dietrich)

Plate 21. Highly explosive eruptions of Mount St. Helens in Oregon, in 1980, devastated the landscape. This view shows prostrate trees knocked down by one of the highly explosive blasts, bordered by an area of trees scorched but still upright, and then by trees that were more or less unaffected. (Photograph by D. R. Crandell)

Plate 22. Pele's hair on pahoehoe (ropy lava) on slope of Kilauea crater in Hawaii Volcanoes National Park. (Photograph by R. V. Dietrich)

Plate 23. Exposure of chalk along the southern coast of Denmark. These biochemical sedimentary rocks are of the same geological age as the well-known chalk cliffs of Dover (England) and Normandy (France). (Photograph by R. V. Dietrich)

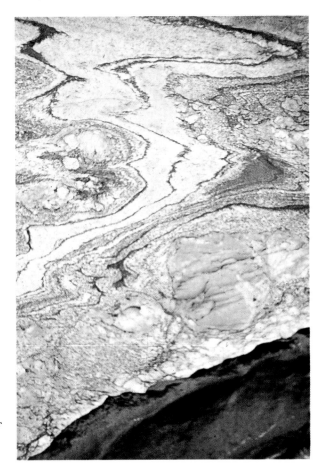

Plate 24. Migmatite, a composite rock comprising light-colored granitic (igneous?) and dark-colored metamorphic rocks. Many rocks of this type represent critical conditions where rocks of, for example, granitic composition become mobile while other rocks remain more or less solid and are metamorphosed. The exposure is in the 30,000 Island district of Georgian Bay, Ontario. (Photograph by Ed Bartram)

Gems, granites, and gravels

Plate 25. Diamond Head, Island of Oahu, Hawaii. (Photograph by R. V. Dietrich)

Plate 26. Devil's Tower, Wyoming. (Photograph by R. Wicander)

Gems, granites, and gravels

Plate 27. A balancing rock in Zimbabwe. (Photograph by R. V. Dietrich)

Plate 28. Ayers Rock, Northern Territory, Australia. (Photograph by R. Wicander)

Plate 29. Chemical weathering: dissolution of exposed corners, edges, and surfaces of a limestone near Brevort, Michigan. Rainwater, which is typically slightly acidic, dissolved calcite, thus rounding off outcrops of these calcite-rich rocks. Corners (with three surfaces of attack) were rounded off more than edges (with only two surfaces of attack), which were weathered more than the relatively flat surfaces. (Photograph by R. V. Dietrich)

Plate 30. Hydration and oxidation were involved in the chemical weathering of this gabbro near Alice, South Africa. Anhydrous feldspar reacted with rainwater to form hydrous clay minerals; anhydrous pyroxene had its ferrous iron react with oxygen in the atmosphere to form ferric iron, which in turn formed the hydrous ferric iron minerals that give the weathering product its reddish brown color. (Photograph by B. J. Skinner)

Plate 31. Laterite in the Ivory Coast. Intense leaching of a relatively flat landscape by warm, tropical rains removes soluble constituents and leaves an insoluble residue of ferric and aluminum hydroxides. The ferric hydroxides are responsible for the red color. (Photograph courtesy of Jean Bahr)

Plate 32. An ancient copper figurine found in the area between the Tigris and Euphrates rivers of Mesopotamia. This statuette, crafted by the Chaldeans, bears an inscription, in Sumerian: "For Bau, the good lady, the daughter of An, the lady of the holy city. His lady, Gudea governor of Lagash the man who built the temple E-Ninnu of the god Ningirsu, built her wall of the holy city." The specimen, dated ~2250 B.C., is in the Yale University Babylonian Collection. (Photograph by W. Sacco)

Plate 33. A smelting operation. The pouring of ingots of molten copper from the smelter at San Manuel, Arizona. (Photograph by B. J. Skinner)

Gems, granites, and gravels

Plate 34. Seafloor chimney, the Juan de Fuca Chimney. *This color pen-and-ink "interpretive drawing" by Calvin Fountain is based on a photomosaic assembled from submarine Alvin's external photos. The seafloor chimney is about 9 m high and 2 m in diameter. A "black smoker" vent issues from the side; tube worms, which average 1 cm × 5 cm, are visible on the chimney wall. (Photograph, which first appeared on the cover of* Geology, *vol. 14, no. 10, courtesy of W. C. Shanks)*

Plate 35. Fine sedimentary layers of a banded iron formation from the Marquette District, Michigan. These alternate layers of cherty silica (gray) and iron minerals (red) were deposited in a marine basin more than 2 billion years ago. Banded iron formations contain the world's largest resources of minable iron ore. (Photograph by H. L. James, courtesy of Economic Geology*)*

Plate 36. Cerussite ($PbCO_3$), like this specimen – NMNH #117514, from the Fluy mine, Arizona – is fairly common in the upper oxidized zone of some galena (PbS) (i.e., lead) deposits. (Photograph by D. Penland, courtesy of Smithsonian Institution)

Gems, granites, and gravels

Plate 37. Smithsonite ($ZnCO_3$) – such as this fine specimen, NMNH #R17868, from the Kelly mine, New Mexico – is a relatively common secondary mineral in oxidized zones of zinc-bearing ore deposits. (Photograph by D. Penland, courtesy of Smithsonian Institution)

Plate 38. Several nonmetallic minerals constitute valuable economic deposits. These sulfur (S) crystals are from Agrigento, Sicily, Italy. (Photograph by W. Sacco)

Gems, granites, and gravels

Plate 39. Other valuable nonmetals: **(a)** *Fluorite* (CaF_2) *from the Huanzala mine, Peru, NMNH #148359.* **(b)** *Barite* ($BaSO_4$) *from Cumberland, England, NMNH #B11543. (Photographs courtesy of Smithsonian Institution)*

Gems, granites, and gravels

Plate 40. Stromatolites formed through actions of cyanobacteria, which are tiny aquatic microorganisms. The location is Shark's Bay, Western Australia. (Photograph by B. J. Skinner)

Plate 41. The larva of the caddis fly builds a case of pebbles and other things, held together by a silk it manufactures to protect its soft body. Its ancestors were building such "homes" at least 200 million years ago. (Photograph by Edward S. Ross, California Academy of Sciences)

Gems, granites, and gravels

Plate 42. Boulders and cobbles are used widely in construction, especially in glaciated areas. This fireplace consists of stones, no two of which are alike, picked up on an island in northern Lake Michigan. (Photograph by R. V. Dietrich)

Plate 43. The Central Library, University City, Mexico. This windowless structure (Juan O'Gorman, architect) was completed in 1952. Each side features a mosaic, the theme of which consists of figures and emblems drawn from cultural antiquities of Mexico. The mosaics comprise natural stones – for example, diverse lavas, sandstones, and onyx (their name for cave formation rocks) – with at least one rock from each of the Mexican states. (Photograph by R. V. Dietrich)

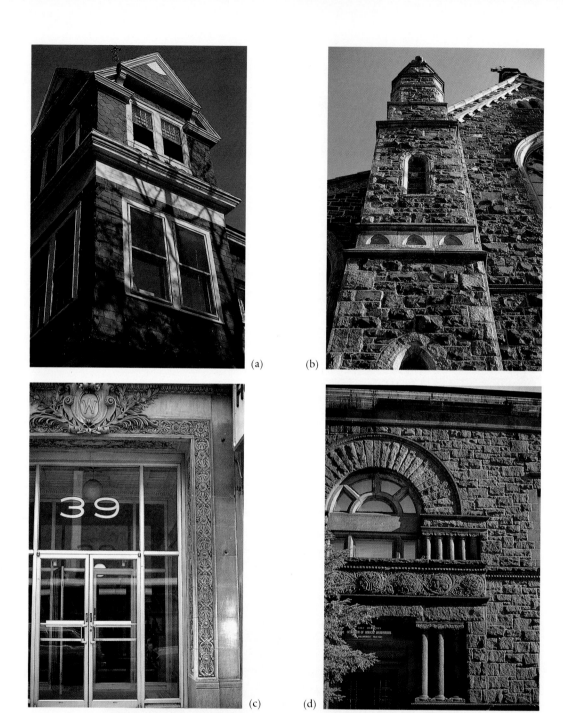

*Plate 44. Many kinds of rocks are "finished" in many ways for use as building stones. The examples shown are all located within a few blocks of each other in downtown New Haven, Connecticut. (**a**) Shingles of red slate, revealing their natural fracture planes, cover this 19th-Century residence on Trumbull Street. (**b**) Rough and chipped blocks of gabbro (dark), edged by rough-cut corner blocks of granite (light) and cut and honed granite (light) window frames, grace the Church of St. Mary (Dominican), the founding site of the Knights of Columbus, on Hillhouse Avenue. (**c**) Carved limestone frames the doorway of this early 20th-Century commercal building on Church Street. (**d**) Arkosic sandstone blocks, some ornately carved, constitute Yale's Museum of Musical Instruments, which is a few doors north of the Church of St. Mary on Hillhouse Avenue. (Photographs by B. J. Skinner)*

Plate 45. Riprap, large rough blocks of quarried rock, may be used for protection and/or as landscape accents. **(a)** These large blocks protect Round Island Lighthouse in the Straits of Mackinac, Michigan – built in 1895, on a near lake-level "reef" – from storm waves. (Photograph by R. V. Dietrich, courtesy of Arnold Transit Company, Mackinac Island, Michigan) **(b)** The riprap around the seventeenth hole and along the shore of its surrounding pond, at the PGA-West TPC® Stadium Golf® Course, La Quinta, California, provide a picturesque setting. (Courtesy of Landmark Communications, La Quinta, California)

Plate 46. Tile has found use, both as structural and ornamental elements, in many cultures for many centuries. This tile roof and the decorative figures (k'uei lung tzu) are on one of the halls in the Forbidden City, Beijing, China. (Photograph by B. J. Skinner)

Plate 47. The Alhambra in Granada, Spain. Built in the 15th Century, the buff and tan panels, such as those above the colorful glazed tile shown in this photograph, are perhaps the world's finest examples of precast plaster. The Arabic characters in Alhambra panels include lines from the Koran, poetry, and statements about historical events. This use of calligraphy for architectural ornamentation is common and in accordance with Moslem tradition, which essentially prohibits the depiction of humans and other animals in visual arts. (Photograph by R. V. Dietrich)

Plate 48. Rose Window in Notre Dame Cathedral, Paris. Several magnificent stained-glass windows like this one were created during the Renaissance. This one was dismantled during World War II and reassembled thereafter. (Photograph by R. V. Dietrich)

Plate 49. Glass-clad, steel-trussed building. This is the 716-foot-high, fifty-nine-floor, First Interstate (originally Allied Bank) Tower in Dallas, Texas (Henry N. Cobb, architect), completed in 1986. A prismatic glass tower that rises from a garden of fountains, waterfalls, and cyprus trees, this edifice has been described as a "lazer-cut gem that never looks the same twice." Above its square footprint, its contained space consists largely of a central parallelepiped. The tower has four vertical faces and four sloping surfaces – two triangular, two rectangular – each pair with a different angle from the vertical. The glass panes, which are set in extruded aluminum frames with a clear anodized coating, are clear and have a highly reflective green coating that photographs blue. (Photograph by Richard Payne, courtesy of I. M. Pei and Partners)

Plate 50. The Swiss Alps near Grindelwald. Many major topographic features, such as the Alps, can be explained on the basis of the Plate Tectonics Hypothesis. As an example, the Alps, where rocks once on the seafloor are now exposed thousands of feet above sea level, are assumed to represent the result of a collision between once separated northern and southern Europe. (Photograph by R. V. Dietrich)

Plate 51. The Grand Canyon of the Colorado. Strata of several diverse sedimentary rocks, ranging in age from more than 600 million years to about 250 million years old and up to nearly a mile thick, are exposed above even more ancient metamorphic and igneous rocks (which represent ancient, weathered and eroded mountain ranges) that are exposed in the lower parts of the canyon. (Photograph by L. E. Andrews)

4

Rocks

A rock is a rock is a rock – sometimes attractive when polished and put on display, but in many cases dull and featureless, or so one might think. But no matter how dull a rock may seem, first appearances can be misleading. Pick up a specimen of one of those seemingly dull rocks and break it so that a fresh, new surface is exposed. Then, look at it closely. The chances are good that a host of tiny mineral grains – possibly randomly oriented, possibly arranged in some sort of rough pattern – will become apparent. The presence of a pattern, or even the lack of one, can tell much about how and where the rock was formed. The way to decipher the story of its formation, however, is to start by studying the individual, constituent mineral grains (Fig. 4.1).

By viewing the grains through a magnifier, each mineral grain may be seen to present either a cleavage or rough fracture surface, which will help identify the individual mineral components. On further inspection, such things as bits of organic matter or fragments of shells (fossils) might be seen to be lodged among the grains. If so, one is seeing the remains of animals or plants that lived long ago and died and then were entombed and preserved among the mineral grains that constitute the rock.

In any case, once one learns how to look at rocks and how to interpret what is seen, it is quickly apparent that rocks are never featureless and are far from dull. In fact, rocks are veritable treasure troves of fascinating information. They are, in fact, the documents that record the events of our Earth's long history. In a sense, rocks are the words of the record, and minerals are the letters that make up those words.

Scientists tend to draw on Greek and Latin word roots when they formulate new names. Those who study rocks are no exception. The old Greek word for a stone or a rock is *petros,* and not surprisingly the scientists who specialize in the study of rocks call their branch of geology *petrology* and refer to themselves as petrologists.

Gems, granites, and gravels

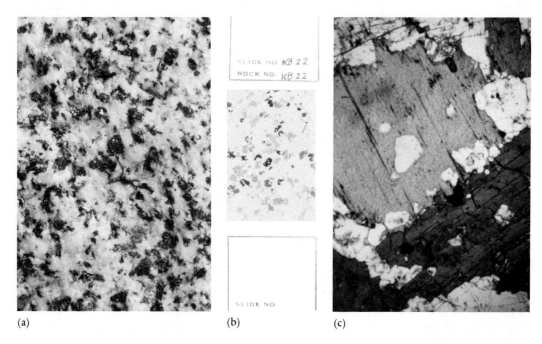

Figure 4.1. The way assemblages of minerals are studied using a petrographic (polarizing) microscope. (a) Hand specimen of granodiorite, an igneous rock that consists largely of quartz, plagioclase feldspar, and a mica called biotite. (b) Thin slice of granodiorite (approx. 24 × 32 mm) that has been glued to a glass slide and then carefully ground down to a thickness of 0.03 mm (approx. 1/1000 of an inch), at which most minerals are transparent. (c) Thin section view showing details of the interrelations that become apparent when one studies the rock with a polarizing microscope.

ROCK: A DEFINITION

Although the work done by petrologists is similar to that done by mineralogists, petrologists tend to emphasize the differences between minerals and rocks by saying that minerals are homogeneous, whereas rocks are heterogeneous. Although this is generally true, the distinction is much too cut-and-dried to be applied to many mineral and rock specimens. The differences may become clear from the following:

Rock is to mineral as forest is to a lonesome pine; as an exquisite tapestry is to a piece of silk thread; as a spring bouquet is to a daisy.... Certain forest-to-tree relationships clarify a number of the diverse aspects of the rock-to-mineral relationships rather well: Although many natural forests contain several species of trees, and most rocks contain several different minerals, some forests are made up mainly of numerous trees of a single species just as some rocks are composed largely of many grains of a single mineral. Also, just as the trees of a forest may be of different shapes and sizes, so may the mineral grains of a rock be of many shapes and sizes. And, although some forests are similar and may be classified together, each is truly unique, just as each rock is unique. (From Dietrich, 1980 [see "Further reading"])

The analogy in this quotation, however, falls short of providing an unambiguous basis for a definition that would embrace all rocks. This is so because it suggests that rocks are composed entirely of minerals, whereas some rocks are made up wholly or partly of natural glass, and a few other rocks – of which coal is an example – consist largely of organic remains called *macerals*. Neither glasses nor macerals fit the definition of a mineral. An inclusive definition for rock is as follows:

A *rock* is a natural solid aggregate composed of mineral grains, glass, macerals, and/or other natural solids.

Inclusion of the word "natural" in the definition means that all manmade materials are excluded. For example, concrete would otherwise fit the definition, but it is man-made and therefore is not a rock.

The term "solid" in the definition is somewhat subjective. Petrologists differ among themselves about what is solid and what is not solid; for example, at what point does a group of sand grains become the rock sandstone? Many rock materials range from loose aggregates, through partially consolidated masses, to coherent rigid masses. Thus, the placing of definitional boundaries is partly subjective, a matter of personal opinion. One "criterion" often mentioned by many geologists is this: If one needs a hammer to break it, it is rock, but if only a shovel is needed, it is not.

The reason for using the designation "mineral grains," rather than just saying "minerals," is that the mineral components of some rocks are innumerable grains of only a single mineral species rather than of two or more different minerals.

The use of the "and/or" emphasizes the fact that a given rock may be made up entirely of mineral grains, entirely of glass, entirely of macerals, or of some combination of mineral grains, glass, and macerals.

The reason for the inclusion of "other natural solids" is discussed in a later subsection.

ROCK COMPONENTS

Common rock-forming minerals

The common rock-forming minerals comprise less than 1 percent of the approximately 3,500 currently known mineral species. All of the common rocks, plus many of the less common rocks, are made up wholly or largely

Table 4.1. *The rock-forming minerals*

(1)	(2)	(3)	(4)
Quartz	**Calcite**	**Biotite**	Hematite
Alkali feldspars	**Dolomite**	**Muscovite**	**Magnetite**
Plagioclase feldspars	Clays	Garnets	Pyrite
Pyroxenes	**Gypsum**	Amphiboles	Limonite
Olivines	Anhydrite	Chlorite	Epidote
Feldspathoids	**Halite**		

Notes: Mineral names given in the plural (i.e., those with a final *s*) are *groups* rather than individual species. Most *igneous* rocks are named on the basis of the presence and relative percentages of one or more of the minerals in column 1. The minerals in column 2 are, along with quartz, the predominant minerals in nearly all of the common *sedimentary* rocks. The minerals in column 3 are, along with quartz, the feldspars, calcite, and dolomite, the chief constituents of most *metamorphic* rocks. The minerals in column 4 are the especially common *accessory* minerals, plus limonite (a common weathering product), that are present in many rocks.

of the eighteen rock-forming minerals that are printed in boldface type on Table 4.1. Thus, along with considerations such as the grain size, the presence or absence of patterns in the distribution or orientation of the grains, and other features mentioned for the different kinds of rocks described later in this chapter, most rocks can be named rather simply on the basis of the presence or absence and percentages of only one or a few of the minerals in Table 4.1.

The skill needed to identify rocks starts with the requirement that one be able to identify these common minerals. Fortunately, the rock-forming minerals are few in number and usually can be distinguished rather easily, even macroscopically. Two tables that will aid one to identify the more common rock-forming minerals are given in Appendix 2.

Natural glasses

Natural glasses are formed whenever silica-rich molten rock material is quenched. As previously mentioned, rapid cooling precludes organization of the constituent atoms into the regular geometric lattices of crystal structures. Instead, it results in the formation of the supercooled liquid we call glass. The apparent rigidity of a glass is merely an expression of its extremely high viscosity.

Natural glasses can be formed in several ways:

fulgurites are produced from rock melt formed when lightning strikes soils, sands, or, in some cases, rocks;

obsidian (Plate 17) and *tachylyte* are formed when magmas with relatively

Figure 4.2. Coal seam in southwestern Virginia. This diagenetic rock is one of the most important sources of energy in the world. (Photograph courtesy of Norfolk & Western Railway)

high and low silica contents, respectively, are quenched – for example, by flowing into water;

pumice is frothlike obsidian;

pele's hair is brown threadlike fibers of tachylyte.

Natural glasses have greasy to vitreous lusters and other characteristics that mimic those of man-made glass. Obsidian, which is by far the most frequently encountered natural glass, is typically medium to dark gray or brown in color and nearly opaque to translucent in large pieces, but transparent in thin slivers. Tachylyte is typically greenish black to black and opaque, even in thin slivers. In addition, tachylyte is readily dissolved in acids, whereas obsidian is not.

Macerals

The macerals comprise a rather large group of organic materials that constitute coal (Fig. 4.2) and portions of several other rocks. The term was first introduced in 1935 by the eminent British coal petrologist Marie Stopes, who in introducing the designation said that "the word 'macerals' will, I hope, be accepted as a pleasantly sounding parallel to the word 'minerals'." Her hope has been fulfilled.

In essence, the term is applied to rock-forming organic units derived by maceration of pieces and products of vegetation, such as woody tissue, spores, and fossil charcoal. There are three chief groups of macerals:

vitrinites are derived from cell-wall material or woody tissue;
liptinites (also called *exinites*) include waxy and resinous materials, such as spores and resins;
inertinites are altered and degraded organic matter – for example, fossil charcoal.

It is often difficult to distinguish macerals from one another by simple examination. Frequently, however, four *coal types* can be identified macroscopically:

vitrain is characterized by a uniform brilliant luster, black color, and conchoidal fracture;
durain ranges from essentially homogeneous to poorly banded, has a dull luster and lead-gray to brownish black color and breaks to give granular surfaces;
clarain consists of dull and glossy layers and tends to break with smooth fractures that are nearly perpendicular to the bedding;
fusain (sometimes called *mineral charcoal*) is typically cellular to porous, has a silky to satiny luster and is dirty to the touch.

Other natural solids

Substances such as opal, asphalt, and amber are other natural solids that occur as minor to major constituents of some rocks.

Opal (Plate 18) is a mineral gel. Gels, like glasses, do not have specific chemical compositions; thus, the composition of opal can only be expressed approximately by the formula $SiO_2 \cdot nH_2O$ (where n is any fraction or number ranging up to about 3.5). The H_2O in the formula is believed to be present in the tiny interstices among the silica spheres that constitute the bulk of the structure (Fig. 4.3).

Asphalt is a black or dark brown bituminous material that is so viscous it seems to be solid. The thick, gummy residue remaining after evaporation of volatiles from petroleum, asphalt consists largely, if not wholly, of carbon and hydrogen.

Amber is an amorphous fossil resin that was exuded from trees and subsequently became hard because of the loss of volatiles. It is of special interest to geologists because it commonly includes insects, pollen, and other representatives of animals and plants that lived and became stuck

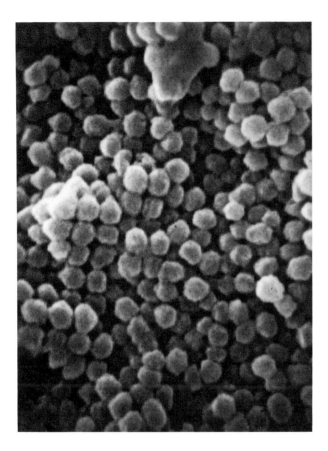

Figure 4.3. Electronmicrograph showing the packing of spheres in a Brazilian opal (28,000×). The play of colors in precious opal (Plate 18) is due to the regular packing of tiny spheres such as those shown. The spheres diffract light rays in the same way that a geometric packing of atoms (which are, of course, much smaller than the silica spheres) diffracts very short wavelength X-rays (see Figure 2.6). (Electronmicrograph by Laura J. Hogan)

in it when its parent resin was formed. In fact, even bubbles of ancient air have been be trapped, thus providing samples of ancient atmospheres.

THE CLASSIFICATION AND NAMING OF ROCKS

Rocks are usually classified into three major families, based on their modes of origin. Three "minor" families are often added to include hybrid or transitional rocks. There also is a seventh catchall group of "other rocks" that includes rocks that do not fit well into any of the six major or minor families.

The three major rock families are the igneous, sedimentary, and metamorphic rocks. The minor families are the pyroclastic rocks, which are transitional between igneous and sedimentary rocks; diagenetic rocks, whose mode of formation includes both sedimentary- and metamorphic-like processes; and migmatites, which are transitional in character between igneous and metamorphic rocks.

The seventh category includes such rocks as the following: veins, such as those in some ore deposits; coherent aggregates of rock materials, such as bauxite (the chief source of aluminum), that are produced as the result of weathering processes; the previously mentioned fulgurites; and the rather exotic rocks produced by the impact of large meteorites and comets on the Earth.

Rocks in each of the categories are named on the basis of features that reflect their modes of formation. For example, igneous rocks are named on the basis of the minerals they contain and their grain sizes, which reflect the composition and speed of cooling of their parent magmas, whereas detrital sedimentary rocks are named on the basis of the sizes of their constituent fragments, generally called clasts, which reflect the character of the agent that transported and deposited them. Nomenclature schemes for all rocks are given in appropriate textbooks. Tables that will aid one to name some of the more common rocks are given in Appendix 2.

ROCK ORIGINS

Rocks, as previously mentioned, are the documents in which the geological history of the Earth is preserved. Consequently, one of the aims of petrologists is to decipher geological history. In particular, petrologists attempt to determine the origins of rocks and to establish the conditions under which they were formed. The motivation for such studies is a combination of scientific curiosity and the anticipation of increased economic returns. Whatever the primary objective, the results of nearly all of these studies add information to the data bank of petrology.

Petrology, the branch of geology specializing in the description, classification, and interpretation of rocks, comprises many kinds of special scientific endeavors – for example, the mapping of rock units in the field; the description of the components, compositions, and textures of rocks, both in the field and in the laboratory; the synthesis of data; the formulation of hypotheses of origin for diverse rocks; and the devising and carrying out of experiments to check critical aspects of those hypotheses.

Petrologists are staunch believers in the utility of the *principle of uni-*

formitarianism, which is one of the most important principles of geology. In essence, the principle holds that modern geological processes are indicative of the kinds of processes that were active in the geological past; to put it on a more scientific basis, the principle holds that the physical and chemical laws of today existed, and thus controlled geological processes, in the past. Geologists have a creed that embodies the principle: The present is the key to the past.

Igneous and pyroclastic rocks

There are few sights in nature as fascinating and awe-inspiring as an incandescent stream of white-hot lava cascading down the flank of a volcano, or a fountain of lava spewing higher than skyscrapers (Plate 19). By contrast, there are few sights so tranquil as the view of Half Dome from Yosemite Valley on a bright moonlit night (Plate 20). And there are few spectacles so disturbing as the devastation around Mount St. Helens (Plate 21), or the town of St. Pierre on the West Indies island of Martinique, where more than 30,000 people were killed during a single eruption of Mount Pelée in 1902. Awe-inspiring, devastating, or tranquil, each of these scenes manifests a kind of igneous or pyroclastic activity that has been repeated many times at many places throughout geological time.

Incandescent lava flows cool and consolidate rather rapidly to form a dark-colored fine-grained rock called basalt; lava fountains yield such rocks as basaltic volcanic bombs and pele's hair (Plate 22). The rock known as Half Dome, the exposed part of which measures approximately 1,450 m (~4,770 ft) in height, is a light-colored coarse-grained granitic rock that was formed from magma that consolidated slowly, well below the earth's surface, and has subsequently been exposed by weathering and erosion of the overlying rocks. The rocks formed during the highly explosive eruptions responsible for the devastation around Mount St. Helens and Mount Pelée are light to medium gray pyroclastic rocks made up largely of fine ash-size fragments of igneous rocks blasted out of volcanic vents.

Each of these observations correlates well with the pertinent facts in the petrology data bank: Both igneous and pyroclastic rocks are derived from magmas. The more rapidly a magma cools, the finer the grain size of the resulting rock (Fig. 4.4). The chemical composition of a magma controls its explosivity and the type of the resulting rock – for example, its mineral composition and color.

As can be seen, all igneous rocks (from the Latin *ignis*, meaning fire) are formed by the consolidation of molten or partly molten rock material,

Gems, granites, and gravels

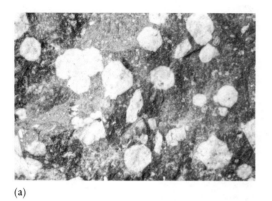
(a)

(b)

Figure 4.4. Igneous and pyroclastic rocks. (a) A porphyry with large crystals of nepheline ($NaAlSiO_4$) within fine-grained matrix, generally called the groundmass. Grain size in igneous rocks is related to cooling rates. Large crystals indicate slow cooling and slow growth of crytals, whereas small crystals indicate cooling so rapid that large crystals did not have time to grow. Porphyries may form when partly crystallized magma, which cooled slowly at depth, is suddenly erupted and its remaining molten portion is cooled rapidly. The sample is from Montana. (b) Tuff, a pyroclastic rock that consists of fragments of igneous rock blasted out of a volcanic vent, deposited as a blanket around the vent, and cemented to form a solid rock. The sample is from Nevada. (Photographs by W. Sacco)

which is generally called *magma* but can also be termed *lava* when it flows out of a volcano onto the land surface. Pyroclastic rocks (from the Greek *pyr,* fire, and *klastos,* broken) are hybrid rocks that are, as the petrologist Chester K. Wentworth once quipped, "igneous on the way up and sedimentary on the way down." That is, they consist of igneous fragments, blasted out during volcanic activities, that then fell to the ground to form sedimentary deposits.

Sedimentary and diagenetic rocks

Anyone who frequents certain stretches of the seashore has witnessed the fact that beaches that are covered by sand during fairly calm weather may be covered with pebbles, cobbles, or even boulders during or following a storm. This reflects the fact that storm waves of relatively high velocity tend to winnow out and carry away sand and small stones, leaving only the larger, heavier stones behind, whereas waves of lower velocity, "calm-weather" waves, can carry sand onto the beaches but do not have enough energy to remove it. That is, the sand remains there because the ensuing offshore flow is too weak to carry it back seaward.

We also know that people who pan for gold are much more likely to be successful when they pan dark-colored sands, which typically consist

Figure 4.5. Sedimentary rocks such as these include two of the common characteristics of rocks of this category: They are layered (stratified) and they are made up of clasts – in this case, sand and gravel. This exposure, called "Turnip Rock," is near the tip of Michigan's "thumb." It is made up of rocks similar to those that once were the basis of a thriving millstone industry at Grindstone City, a few miles to the northeast. (Photograph by Marion I. Whitney)

largely of dense minerals, than when they pan the more common light-colored, typically less dense sands. This depends on the fact that currents with velocities great enough to transport (and deposit) dense mineral grains can also carry grains of gold, which is the most dense mineral, whereas currents with lower velocities, high enough to transport only light, low-density sands, can carry only extremely small, if any, particles of gold.

These basic facts explain two common characteristics of loose sediments formed from fragments transported in flowing water (as well as in other fluids, including air) and of the sedimentary rocks formed from them:

1. The constituent fragments tend to be well sorted.
2. The sediments and rocks formed from them are typically layered – that is, they are *stratified* (Fig. 4.5).

The individual fragments are often called *clasts,* and deposits made up largely of these fragments are generally termed *clastic sediments*.

Flowing water can also transport dissolved rock materials in solution. When the dissolved matter becomes sufficiently concentrated, minerals are precipitated, thus forming another kind of sedimentary rock. The spring deposits at Yellowstone National Park and the stalactites and other depositional formations – which, as a group, are called *speleothems* – that

(a) (b)

Figure 4.6. Another common characteristic of sedimentary rocks is the presence of fossils. (a) Shell fragments like those accumulating today have been cemented together to form this less-than-1-million-year-old rock, which is often called coquina. (Photograph by B. J. Skinner) (b) Approximately 370-million-year-old fossils are included in this fossiliferous limestone. (Courtesy of Smithsonian Institution)

occur in many caves are obvious examples of such precipitates. The Earth's widespread salt deposits, such as those that occur in the ~415-million-year-old Silurian formations, which are mined for halite (salt) near Syracuse, New York, and in the region around Detroit, Michigan, and Windsor, Ontario, comprise even greater volumes of this kind of sedimentary rock. All such precipitates are referred to as *chemical sediments*.

In some cases the precipitation of minerals from solution is not a strictly inorganic chemical process; rather, it is promoted by the activities of one or more organisms. Accumulations formed as a result of these activities are properly called *biochemical sediments*.

In a broad sense, shells and other mineral components of dead plants and animals that are deposited with sediments are examples of biochemical precipitation. It is these hard parts that are most likely to become preserved as fossils (Fig. 4.6), which are characteristic of many sedimentary rocks. One of the more amazing biochemical sedimentary rocks is chalk: The ~70-million-year-old chalk formations exposed in the famous White Cliffs of Dover in England and comprising the picturesque stacks and other nearshore features along the southern coast of Denmark and the northern coast of France consist almost entirely of fossils (Plate 23). In fact, a specimen of chalk no larger than one's thumb contains literally hundreds of thousands of microscopic exoskeletons and other biochemically precipitated parts of marine organisms.

The particles in a loose sediment, no matter how formed, must become *lithified* in order to be turned into rocks. Lithification comes about as a result of one or more of a group of processes termed *diagenesis*. These processes include such things as the solution and reprecipitation of mineral grains by groundwater or seawater, deposition of new material introduced in solution in groundwater or seawater, and the replacement of original mineral grains by chemical reactions with groundwater or seawater. They are the same kinds of processes that are involved in metamorphism; the difference is that diagenesis takes place in sediments at low temperatures and pressures, whereas metamorphism occurs in solid rocks subjected to elevated temperatures and pressures.

Sediments are sometimes changed to such a degree that geologists refer to the resulting rock products as diagenetic rocks. Many dolostones, which are rocks that consist largely of the mineral dolomite [$CaMg(CO_3)_2$], and many cherts, which are rocks made up predominantly of extremely fine grained quartz (SiO_2), are diagenetic rocks. For both, the typical precursor sediments were made up largely of either calcite or aragonite, which are polymorphs of calcium carbonate ($CaCO_3$). That is to say, the dolomite and quartz were formed by diagenetic reaction or replacement of the original calcite or aragonite (Fig. 4.7). Several other rocks, including all of the diverse kinds of coal, which were originally swamp and bog materials, also owe at least part of their current characteristics to diagenetic processes.

As is evident, then, sedimentary rocks (from the Latin *sedimentum*, settling, from *sedere*, to sink or sit down) have two principal modes of origin:

1. by lithification of aggregates of clasts, such as sands and gravels, that have been transported and deposited on the surface of the Earth, and
2. by chemical or biochemical precipitation of minerals from aqueous solutions either on or near the surface of the Earth.

Most sedimentary rocks are layered (stratified), and many contain fossils. Diagenesis (from the Greek *dia*, through, and *genesis*, origin) is the term applied to chemical and physical modifications and transformations other than those attributable to metamorphism or weathering. Diagenetic processes – which include solution, deposition, replacement, and recrystallization – take place in loose sediments and frequently contribute to their becoming coherent rocks.

Sedimentary and diagenetic rocks provide many clues about the environments that existed when their precursor sediments were deposited. For example, often it is possible to say whether a sediment was deposited in a marine or nonmarine body of water, to estimate the general depth

Gems, granites, and gravels

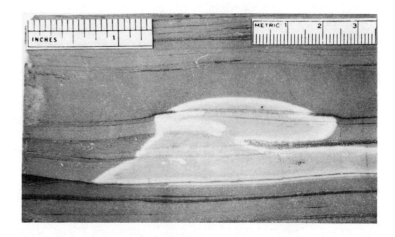

Figure 4.7. Diagenetic chert (white) within dolostone (gray), both of which replaced a precursor calcium carbonate sediment. Note that the original sedimentary bedding – that is, the essentially horizontal layers and lamellae – has been preserved. Specimen is from exposures near Luster's Gate, Montgomery County, Virginia. (Photograph by R. V. Dietrich)

of the water where a sediment was deposited, and to deduce the topographic character and probable climate over the landmass from which the clastic fragments or solutions were derived. In addition, often one can decipher both the processes responsible for lithification and the conditions that existed during lithification.

Metamorphic rocks and migmatites

Metamorphic rocks and migmatites make up more of the Earth's continental crust than the combined total of all the other kinds of rocks. Although many of these metamorphic rocks and migmatites are below the surface, hidden from view, fine exposures occur in many places – for example, here and there on the ancient Precambrian shields and also in several mountain masses, such as the Appalachians and the Rockies of North America, the Andes of South America, the Alps and the Pyrenees of Europe, the Atlas Mountains of northwestern Africa, and the Himalayas and Urals of Asia.

The basic concept of metamorphism was first expressed by James Hutton, often called the father of modern geology. Hutton, an 18th-Century Scottish physician, farmer, and naturalist, wrote:

If, in examining our land, we shall find a mass of matter which had been evidently formed originally in the ordinary manner of stratification, but which is now extremely distorted in its structure and displaced in its position, . . . which is also extremely consolidated in its mass and variously changed in its composition, . . . which, therefore, has the marks of its original or marine composition extremely obliterated, and many subsequent veins of melted mineral matter interjected, we should then have reason to suppose that here were masses of matter which, though not different in their origin from those that are gradually deposited at the bottom of the ocean, have been more acted upon by subterranean heat and the expanding power, that is to say, have been changed in a greater degree by the operations of the mineral kingdom. (From Geike, 1897 [see "Further reading"])

The group of processes Hutton was describing are now lumped together and called metamorphism. The term is, in fact, now applied to all physical and chemical transformations of rock that occur at temperatures and/or pressures higher than those under which diagenesis takes place and lower than those under which the given rock would melt.

Metamorphism produces changes in textures, in mineral contents, and even in overall chemical compositions of some rocks. The controls of metamorphism are pressure, temperature, and chemical environment. The pressures involved may be either hydrostatic (= confining) or directed (= differential stress), or a combination of both. Numerically, the pressure range over which metamorphism occurs is from one atmosphere up to several thousand atmospheres. The temperatures, as just mentioned, may range up to the melting temperature of the given rock – that is, from about 150°C up to about 850°C. The chemical environment may include only the elements in the original rock, or it may also involve additions and/or subtractions of constituents carried by through-flowing fluids.

A rock will tend to be metamorphosed whenever it is subjected to changes in temperature and/or pressure and/or chemical environment if the new conditions differ markedly from the conditions under which the rock originated. Given sufficient time under the changed conditions, the rock will respond and become metamorphosed. The metamorphism will be manifested by adjustments in texture and/or mineralogical makeup. The processes are analogous to those that occur during baking:

When a mixture of flour, salt, sugar, yeast, and water is baked in an oven, the high temperature causes a series of chemical reactions – new compounds grow and the final result is a loaf of bread. When rocks are heated, new minerals grow and the final result is a metamorphic rock. In the case of the rocks, the source of heat is the Earth's internal heat. Rock can be heated simply by burial, or by a nearby igneous intrusion. But, burial and the process of intrusion can also cause a change in pressure [and in chemical environment]. Therefore, whatever the cause of the heating, metamorphism can rarely be considered to be entirely due to the rise in temperature. The combined effects of changing temperature and pressure [and chemical composition] must be considered together. (From Skinner and Porter, 1987 [see "Further reading"])

Metamorphic rocks are found (1) in zones of intense deformation, such as fracture zones in the Earth's crust, where large masses of rocks have been ground against each other as they have been broken and offset, (2) near bodies of igneous rock, where heat and vapors given off by an intruded magma have permeated the adjacent rocks, and (3) in the cores of mountain belts, where large volumes of buried rock have become heated and subjected to high pressures. These three situations are usually referred to as dislocation metamorphism, contact metamorphism, and regional (or dynamothermal) metamorphism, respectively. The designation "regional" directs attention to the fact that metamorphic rocks of the last type typically underlie extremely large areas, such as the "crystalline Appalachians" that extend from Georgia, in the southeastern United States, northeastward through the Maritime Provinces of eastern Canada and appear to have originally included rocks of similar age and nature that underlie eastern Greenland and parts of Scotland, Ireland, Norway, and Sweden.

As a response to growth or recrystallization under pressure, the new minerals in many metamorphic rocks grow with preferred orientations. Examples include rocks such as gneiss (pronounced \nīs\) schist, phyllite \fill′īte\, and slate, all of which contain platy and/or tabular minerals with their short dimensions nearly parallel. The texture produced by the parallelism of platy or tabular minerals, which is called *foliation* (from the Latin *foliat,* leafy, and the suffix *-ion,* condition), is one of the most obvious characteristics of many metamorphic rocks (Fig. 4.8). Even some metamorphic marbles and quartzites that exhibit no obvious foliation have preferred orientations of the crystal structures of their constituent minerals that become evident as a result of microscopic studies.

Migmatites are transitional between metamorphic and igneous rocks, at least in general appearance. Typical migmatites consist of two fairly distinct constituents: a dark-colored, foliated metamorphic rock and a light-colored rock that is, or at least resembles, a granitic igneous rock (Plate 24). A typical migmatite looks as if its light-colored constituent was at one time more mobile than its dark-colored constituent. Many flamboyant to psychedelicate-appearing migmatites occur in zones between distinctly igneous and metamorphic rock units. It would appear that the conditions of temperature and pressure under which most migmatites were formed were just below those required for extensive melting.

Rocks

Figure 4.8. A foliated metamorphic rock: gneiss. Rocks like this are formed under conditions of high temperature and elevated pressures that exist, for example, at great depths. The exposure is near the Blue Ridge Parkway in southwestern Floyd County, Virginia. (Photograph by R. V. Dietrich)

Therefore, as we have seen, metamorphic rocks (Greek *meta*, change or trans, and *morphē*, form) are formed as the result of the transformation of preexisting rocks. Foliation is a typical characteristic of many metamorphic rocks. Migmatites (Greek *migma*, mixture) are intimate mixtures of metamorphic and igneous or igneous-appearing rocks. Some petrologists refer to migmatites as ultrametamorphic, which emphasizes their niche between metamorphic and igneous rocks.

Other rocks

Moondust probably sounds more like a song title than a rock. Nonetheless, up until a few years ago, one of the favored hypotheses for the formation of the glassy rocks called tektites had them coming from the moon. It was held that they had been formed by lunar volcanism, had escaped the moon's gravitational field, had been captured by the Earth's gravitational field, and had fallen to the Earth's surface. We now know that hypothesis to be wrong. Instead, tektites are formed when comets, meteorites, or other large extraterrestrial masses hit the Earth. We also know that impacts between large extraterrestrial masses and the Earth have resulted in the formation of yet another group of rocks: those usually referred to as impactites. All of these – tektites, meteorites, and impactites

Gems, granites, and gravels

ROCKS ON DISPLAY

Large rock exposures, such as the Grand Canyon of the Colorado River and Half Dome in Yosemite National Park, are among the most visited tourist attractions on Earth. In addition, some rock exposures have become the subjects of widely acclaimed works of art – including paintings, photographs, and music – and a few have been the focus of international advertising campaigns.

As might be expected, nearly all of the widely known field exposures of those special rocks to which tourists are attracted were first noticed and named because of their shapes, rather than because of the makeup of their constituent rocks. A few, however, owe their fame to legend or to historical coincidence – for example, some of the rocks chronicled by the brothers Grimm, and the loose boulder, called Plymouth Rock, on which the Pilgrim Fathers are said to have landed.

In the column titled "Rock Chips" in **Rocks & Minerals** (Titamgim, 1984 [see "Further reading"]), several of the world's most famous rock exposures are described and listed as follows:

Some named "rocks" are known only to people familiar with the vicinity in which they occur – for example, there is a "Pulpit Rock" near the town where I grew up, which I suspect is known only by a few people who live in that area. At the other extreme, there are several "rocks" that are known so widely that they have become tourist meccas, are depicted on postage stamps....

The following are widely known "rocks." ... The predominant kind of constituent rock or rocks and the location are given for each. Those preceded by an asterisk (*) have been pictured on one or more postage stamps.

***Rock of Gibraltar** – cavernous limestone capped by shale, on the north side of the Strait of Gibraltar between the Mediterranean Sea and the Atlantic Ocean.

***Diamond Head** – the remnants of a basaltic volcanic vent, near Honolulu on the Island of Oahu, Hawaii [Plate 25].

***Old Man of the Mountains** (also called "The Great Stone Face") – granite (plus supporting concrete and guy wires), near Franconia Notch, New Hampshire.

***Devil's Tower** – a basaltic volcanic neck at Devil's Tower National Monument, in northeastern Wyoming [Plate 26].

Giant's Causeway – a columnar-jointed basalt on the northern coast of Antrim, Northern Ireland.

***Sugar Loaf** – a granite porphyry dome in Guanabara Bay, Rio de Janeiro, Brazil.

***Half Dome** and ***El Capitan** – granodiorite mountains in Yosemite National Park, east-central California.

Stone Mountain – a granite dome near Atlanta, Georgia.

Bartolomé – a volcanic tuff "spire" at the west end of James (Santiago) Island of the Galapagos Islands.

Percé Rock – a sea-eroded, pierced rock, limestone mass off the eastern end of the Gaspé Peninsula, Québec.

Balancing rocks – there are several of these in the world, some of which have names. Among the better known are those in the Garden of the Gods, near Colorado Springs, Colorado; those near Dartmoor in southwestern England; those near Salisbury and in the ***Rhodes-Matopos National Park**, Zimbabwe (formerly Southern Rhodesia) [Plate 27]; and the Tandil Rocks of Argentina. All of these so-called balancing rocks are residual boulders, and many are granitic.

Ayers Rock – a red sandstone dome in the outback in central Australia [Plate 28].

Chimney Rock – a sandstone "spire" atop a shale base, near Bridgeport, central western Nebraska.

Blowing Rock – a granitic gneiss, frequently windblown, mountain top near Boone, North Carolina.

And, there are, of course, many, many more....

We direct attention to the following:

the numerous caves and their interesting cave formations (stalactites, stalagmites, and other speleothems) that are present here and there around the world – for example, those dotting the folded Appalachians in Pennsylvania and Virginia, and the Carlsbad Caverns in New Mexico, one of the most visited of the United States' national parks,

the diverse natural bridges, such as the one in Rockbridge County, Virginia, which presumably was surveyed by George Washington, and Rainbow Bridge in Natural Bridges National Monument, in the four-corners area of Utah, Colorado, New Mexico, and Arizona, and

the rock profiles of Bryce Canyon, Utah, and Chiricahua Canyon, Arizona, where with a little imagination one can "see" just about anybody or anything sculpted into the rocks by natural erosion processes.

– are examples of rocks usually relegated to the leftover category of "other rocks."

TEKTITES. Tektites are masses of natural glass, ranging up to about 15 cm (~6 in.) in greatest dimension, that have shapes and surface features that indicate they were formed as the result of quenching under the influence of aerodynamic forces (Fig. 4.9). We are now rather sure that they represent rock materials that were melted and splattered into the air when relatively large extraterrestrial masses collided with the Earth. Tektites typically occur in strewn groups comprising innumerable individuals. Most of them have been named for the areas where they have been found – for example, there are australites, indochinites, philippinites, and moldavites (from the Moldau River Valley in Bohemia).

METEORITES. Apparently because they have "fallen from heaven," meteorites have been venerated, if not actually worshipped, for centuries. For example, in the Bible (Acts 19:35), the temple in the city of the Ephesians is noted as the "keeper of...the sacred stone that fell from the sky." Meteorites are indeed from outer space: Many of them have apparently come from the asteroid belt between Mars and Jupiter; those from "showers," however, are thought more likely to represent fragments of comets. There are three chief kinds of meteorites:

1. those made up almost completely of metallic iron–nickel alloys ("iron meteorites") (Fig. 4.10),
2. those made up almost entirely of nonmetallic minerals ("stony meteorites"), and
3. a transitional group ("stony-iron meteorites").

IMPACTITES. Impactites are rock materials that have been shattered, pulverized, and in some cases partially melted as a result of impact be-

Gems, granites, and gravels

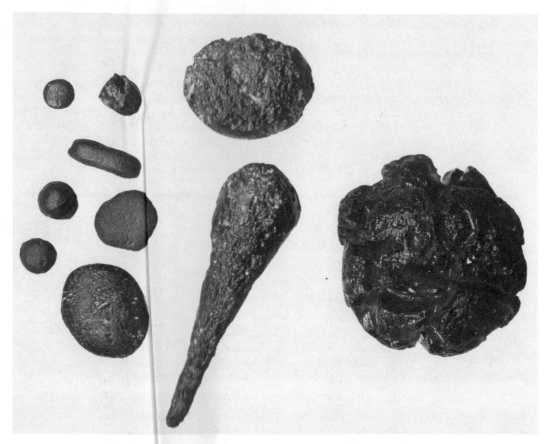

Figure 4.9. Tektites, small glassy objects, shown above at natural size, are believed to form as a result of large meteorite impacts on the Earth. The impact creates such high temperatures that small blobs of molten soil or rock are splashed into space. The regular shapes rise from aerodynamic sculpturing as the molten fragments fly through the atmosphere. These samples, which came from Australia and southeastern Asia, were formed by an impact 700,000 years ago. (Photograph courtesy of Peabody Museum, Yale University)

tween high-velocity meteorites and the Earth. These rocks, sometimes attributed to "shock metamorphism," are the rocks in which the high-density forms of SiO_2 – for example, coesite (see Fig. 3.6) – have been found in nature.

Economically important members of this other-rocks category are treated in the following two chapters. They include such things as laterites and many kinds of ores.

FURTHER READING

The American Association of Petroleum Geologists (P.O. Box 979, Tulsa, Oklahoma 74101) publishes geological highway maps for different regions of North America. These maps (with remarks on the sheets) are excellent "travel companions" for those interested in the geology and geological features they can view along the way.

Berry, L. G., and Mason, B., 1983, *Mineralogy: Concepts, Descriptions, Determinations* (2nd edition by R. V. Dietrich). W. H. Freeman, San Francisco, 561p.
A summary about natural glasses and macerals is given in an Appendix in this book.

Decker, R. W., Wright, T. L., and Stauffer, P. H., 1987, *Volcanism in Hawaii*. U.S. Geological Survey, Professional Paper 1350, 1667p. (2 volumes).
One of the best-illustrated and most authoritative works ever published on volcanism. It was written in celebration of seventy-five years of activity at the Hawaiian Volcano Observatory.

Dietrich, R. V., 1989, *Stones: Their Collection, Identification, and Uses* (2nd edition). Geoscience Press, Prescott, Arizona, 191p.
The quotation is from page 6 of this paperback, which is written for nonprofessionals.

Dietrich, R. V., and Skinner, B. J., 1979, *Rocks and Rock Minerals*. Wiley, New York, 319p.
This basic text gives information about how to name and classify rocks and also an introduction to origins of diverse rocks. It includes tables to aid one to identify both minerals and rocks.

Dictrich, R. V., and Wicander, E. R., 1983, *Minerals, Rocks, and Fossils*. Wiley, New York, 212p.
This inexpensive paperback "self-teaching guide" is for those who may want to collect minerals, rocks, and/or fossils.

Ehlers, E. G., and Blatt, H., 1982, *Petrology: Igneous, Sedimentary, and Metamorphic*. W. H. Freeman, San Francisco, 732p.
An intermediate-level text covering the mineral assemblages, textures, and origins of rocks.

Geike, A., 1897, *The Founders of Geology* (2nd edition). Macmillan, London, 486p.
The quotation from James Hutton, Theory of the Earth, Vol. 1, pp. 275–6, is taken from this interesting book about Hutton and several other early geologists.

Figure 4.10 (facing). An iron meteorite from Staunton, Virginia. This iron meteorite has been cut, polished, and etched to display the internal structure. The black, rounded object is troilite (FeS). The interlocking laths are two different kinds of iron–nickel alloys. The latticelike pattern, called Widmanstätten texture, forms as the result of slow cooling over millions of years. (Photograph courtesy of Smithsonian Institution)

Harris, D. V., and Kiver, E. P., 1985, *The Geologic Story of the National Parks and Monuments* (4th edition). Wiley, New York, 464p.
Includes easy-to-read, well-illustrated descriptions of the geological features of most of the national parks and several of the national monuments.

Hutton: See the Geike entry.

Lipman, P. W., and Mullineaux, D. R. (editors), 1981, *The 1980 Eruptions of Mount St. Helens, Washington*. U.S. Geological Survey, Professional Paper 1250, 844p.
A well-illustrated and documented history of the eruptions and resulting deposits and their general effects on such diverse things as quality of rivers and soils in the region around Mount St. Helens.

Perret, F. A., 1937, *The Eruptions of Mt. Pelée 1929–1932*. Carnegie Institution of Washington (D.C.), Publication 458, 126p.
The fine photographs (for their time) and the humanistic touches in the narrative by the author, who essentially "lived with" the volcano during the 1929–32 episodes of this highly explosive volcano, make for extremely interesting reading.

Skinner, B. J., and Porter, S. C., 1987, *Physical Geology*. Wiley, New York, 750p.
An up-to-date, extremely well illustrated basic textbook of geology.

Tennissen, A. C., 1983, *Nature of Earth Materials* (2nd edition). Prentice-Hall, Englewood Cliffs, New Jersey, 415p.
Another text about minerals and rocks, their naming and classification.

Titamgim, R. D., 1984, Rock Chips (Question 9). *Rocks & Minerals*, 59, 237.
The quotation about the field exposures of well-known "rocks" is from this column.

PERIODICALS

Journal of Petrology. A bimonthly professional journal that deals chiefly with igneous and metamorphic petrology. Oxford University Press, Walton St., Oxford OX2 6DP, United Kingdom.

Journal of Sedimentary Petrology. The bimonthly professional publication of the Society of Economic Paleontologists and Mineralogists, P.O. Box 4756, Tulsa, Oklahoma 74159–0756.

5

Soils, dusts, and muds

Our seemingly fixed and stable Earth is anything but – indeed, it is constantly changing, ever in motion. Nearly everyone is aware of the sudden, dramatic movements of the ground that accompany big earthquakes. Even if they have not experienced an earthquake, most people know about them because newspapers and television newscasts carry stories of the horrors confronting those who do experience high-magnitude quakes. On the other hand, few people realize that no matter where they are, the ground beneath their feet is in relentless, restless movement: rising or falling, tilting, or moving sideways. Granted, most of the movements are so small and usually so slow that even the most sensitive instruments can barely detect them. Nevertheless, we know they are taking place, and we know that over millions of years even these movements, tiny as they are, can add up to gigantic changes.

This is important, because even though a million years is long on the time scale of human history, it is short on the time scale of the Earth's geological history. Indeed, we now know that in the geological past, seascapes have very slowly risen to become mountains, mountains have been eroded away, and in some places mountainous areas have even been depressed and eventually inundated by inland seas. We also know that through the geological ages, deeply buried rocks have been elevated and exposed at the surface and, once raised, have been broken down by natural actions and reactions, such as those caused by rainwater, by the freezing of water, and by plant and animal activities.

The breakdown processes are collectively called *weathering*. The breakdown products include the soil, dust, mud, and other debris that coat the rocky outer surface of the Earth like a loose-fitting, moth-eaten cape. Collectively, all the debris resulting from weathering is called the *regolith,* a term from the Greek word *rhegos,* meaning blanket, and the old English word *lith,* meaning rock – hence *blanket rock.*

Gems, granites, and gravels

(a)

(b)

Figure 5.1. Chemical weathering affects different rocks in different ways. The general styles and locations of the monuments bearing these inscriptions indicate that they were put in place and engraved at about the time of the dates shown. The two monuments are within a few feet of each other and face the same way in a cemetery in central Virginia. Both inscriptions were carved into polished surfaces. The granite (a), which resists chemical weathering under the climatic conditions within the area, retains a good polish and definition of its engraved numbers; the marble (b) has lost the polish on most of its surfaces, as well as the definition of its numbers, because of chemical weathering. (Photographs by R. V. Dietrich)

WEATHERING

We usually think of rocks as hard and durable, as substances that can resist the ravages of time. A visit to an old cemetery in Europe or eastern North America, however, will soon show the error in such a belief. Gravestones, particularly those made of marble, can be seen to be so extensively weathered that they can hardly be read, even though they have been standing for only a couple of hundred years (Fig. 5.1).

There are two distinct kinds of processes that cause weathering:

Mechanical weathering – also termed physical weathering or disintegration – includes the processes that reduce solid rock to loose mixtures of particles as a result of fragmentation.

Chemical weathering – also called decomposition – embraces the chemical changes that occur when rocks are altered by rain, by the atmosphere, or in response to biochemical changes associated with the growth, death, and decay of plants and animals.

The most important mechanical weathering process involves disaggregation by *frost wedging*. When water fills fractures and other openings in a rock and then freezes to ice, there is a 9 percent increase in volume. This increase can exert enough force to break up the containing rock. A similar process occurs when water evaporates and crystals of a salt are deposited within a rock fracture; the growing crystals can sometimes exert forces sufficient to widen cracks and eventually break rocks apart. In addition, roots of trees (Fig. 5.2), burrowing animals, and even the rapid heating of rock by a brushfire can cause mechanical disaggregation.

The three principal processes grouped under the term chemical weathering are dissolution, hydrolysis, and oxidation. Each of these processes involves changes in one or more of the mineral constituents of the rock material being weathered. New minerals are frequently formed.

The first process, *dissolution* (also termed *leaching*), is simply the removal of soluble matter by rainwater, groundwater, or, rarely, dew. Calcite ($CaCO_3$) dissolves readily in rainwater because most rainwater is slightly acidic. Thus, rain falling on limestone and marble – which are made up largely of calcite – will dissolve those rocks, albeit usually rather slowly (Plate 29). Several other minerals, such as dolomite and gypsum, and the rocks that contain them are also subject to ready dissolution.

The second process of chemical weathering, *hydrolysis* (or *hydration*), is a process whereby certain ions present in minerals – for example, potassium (K^{+1}) or calcium (Ca^{+2}) – are replaced by hydrogen ions (H^{+1}) that are present in rainwater or groundwater. One especially

Gems, granites, and gravels

Figure 5.2. Physical weathering. The joint block beside the tree, near the center of this view, weighs more than five tons. Its movement away from the escarpment was apparently caused by wedging in response to the growth of the tree, a hard maple, possibly abetted by frost action. The location is the town of Macomb, St. Lawrence County, New York. (Photograph by R. V. Dietrich)

important hydrolysis reaction involves the alteration of potassium feldspar ($KAlSi_3O_8$), an abundant mineral in granites, to kaolinite [$Al_4Si_4O_{10}(OH)_8$], which is one of the clay minerals. The hydrolysis of potassium feldspar is shown in the following chemical reaction:

$$4KAlSi_3O_8 + 4H^{+1} + 2H_2O \rightarrow 4K^{+1} + Al_4Si_4O_{10}(OH)_8 + 8SiO_2$$

potassium + hydrogen + water → potassium + kaolinite + quartz
 feldspar   ion   ion

The potassium ion that is released goes into solution in the water, and much of it eventually becomes a nutrient for plants.

Oxidation, the third chemical process involved in weathering, is a process whereby certain chemical elements lose electrons – that is, they change their ionic states. (The reverse process, whereby they gain electrons, is termed *reduction.*) Iron is a good example. In nature, iron (Fe) occurs in three common ionic states:

In the *metallic* state, it is uncharged (Fe^0).
In the *ferrous* state, it is doubly charged (Fe^{+2}).
In the *ferric* state, it is triply charged (Fe^{+3}).

In the presence of oxygen, the ferric state (Fe^{+3}) is the stable state. Thus, when minerals such as biotite, olivine, and pyrite, which contain iron in the ferrous state (Fe^{+2}), are exposed to the atmosphere, they are subject to chemical weathering because their ferrous iron tends to be oxidized to ferric iron (Plate 30). Goethite [FeO(OH)] is probably the most common ferric iron compound that forms as a result of the change of Fe^{+2} to Fe^{+3}. Goethite is, in fact, the reddish to yellowish brown mineral that is familiar to all of us as *iron rust*.

Mechanical and chemical weathering processes usually proceed in concert. For example, when mechanical weathering breaks rocks into smaller and smaller pieces, more surface area is exposed, and thus chemical weathering is hastened. The smaller pieces are quickly altered to the clay minerals, goethite, and other minerals that make up most soils.

Usually, either mechanical weathering or chemical weathering predominates within a given region. Mechanical weathering tends to be predominant in cold regions where water often freezes and thaws; chemical weathering is more prevalent in temperate and tropical regions where there is moderate to high rainfall. In any case, the net result is a regolith that covers a large percentage of the land surfaces the world over.

SOILS

Chemical and mechanical weathering of rock is the first step in the formation of soil. The regolith, however, is called soil only if it can support rooted plants. Two things are necessary:

1. The decomposition of minerals must proceed to the point that potassium and other nutrient elements are present in sufficient quantities for plants to draw them upward through their roots.
2. Organic matter (humus) derived from the decay of plant matter and bacterial actions must be present.

That is to say, plants grow in soil because they are nourished there by nutrient chemical elements released during weathering and by decayed organic matter.

Thus, as plants grow and die and add more humus to the regolith, a soil gradually evolves. Depending on factors such as rainfall, temperature, and the extent of plant cover, a mature soil may take as little as a few hundred years or as long as hundreds of thousands of years to develop.

Gems, granites, and gravels

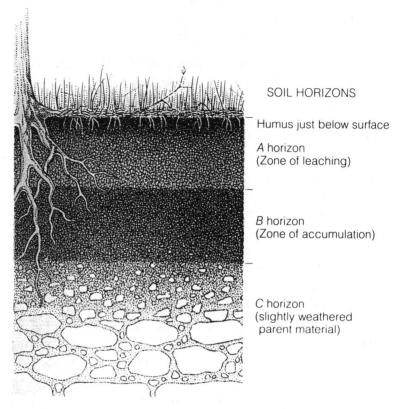

Figure 5.3. Soil profile. Horizons in a typical soil profile, grading upward from fresh rock, through earthy regolith, to humus. The boundaries between horizons are gradual, not sharp.

Whatever the time required, all mature soils consist of a distinctive sequence of layers that can be observed if one digs a pit and looks at the profile.

A *soil profile* (Fig. 5.3) is a succession of distinct horizons with a humus-rich layer at the surface and unaltered rock material at its base. A typical well-developed profile comprises three distinct layers or zones – the A, B, and C layers. The *A layer*, which includes the humus-rich horizon, is the layer from which most soluble constituents have been leached by penetrating rainfall. The *B layer*, below, is the zone in which the leached nutrients tend to accumulate; iron is commonly present in the ferric state, and thus the typical B horizon has a reddish color. The B layer is penetrated by most plant roots; thus, they reach the soil nutrients – for example, potassium, phosphorus, sulfur, and nitrogen – that are accumulated there. The *C layer* consists in part of degraded and weathered parent rock material. The boundaries between the layers are not static; instead, they migrate slowly downward. Soil is, after all, the product of a dynamic system.

Table 5.1. *The ten soil orders of the U.S. soil classification scheme*

Soil order	Meaning of the name	Most important characteristics
Alfisol	Pedalfer soil	Clay-rich B zone covered by thin A zone; in places, a thin, gray-colored layer (A-2) separates A and B; common in humid midlatitudes.
Aridisol	Arid soil	Thin A and B zones and an accumulation of calcium carbonate at the top of the C zone; common in dry climates.
Entisol	Recent soil	Very young, immature soil; thin A zone may be present directly on a C zone.
Histosol	Organic-rich soil	Peaty soil that contains a lot of organic matter; most common in cool, moist climates.
Inceptisol	Young soil	Poorly developed soil with an A zone and a B zone that lacks clay enrichment.
Mollisol	Soft soil	Grassland soil with a thick A zone rich in organic matter and a B zone enriched in clay; calcium carbonate horizon may be present.
Oxisol	Oxide soil	An infertile soil with a highly leached A zone over an extremely weathered and deep B zone.
Spodosol	Ashy soil	Acid soil of a cool forest; highly organic-rich surface horizon and an iron-rich B horizon.
Ultisol	Ultimate soil	Strongly weathered soil with an A zone over a highly weathered B zone; found in tropical and subtropical climates.
Vertisol	Inverted soil	Old soil that has a high content of clay minerals that expand and contract as the moisture content changes.

Source: Adapted from U.S. Soil Conservation Service (1975).

The most important minerals in soils are clay minerals, such as kaolinite and montmorillonite, quartz, calcite, and goethite. The relative abundance of the different minerals in a given soil and the thicknesses of the horizons of the soil profile are determined by a complex interplay between such factors as the parent rock material, the climate, the local topography, time, the soil organisms present, and the vegetation cover.

Soil scientists have found that it is essentially impossible to quantify all aspects of the soil-forming processes into a consistent soil classification scheme. Therefore, they have developed a classification that is based primarily on physical properties. Ten soil orders are recognized in the classification scheme that is widely used in the United States today (Fig. 5.4). The ten orders and their main characteristics are listed in Table 5.1. Two soils of special interest to the petrologist are laterites and caliches.

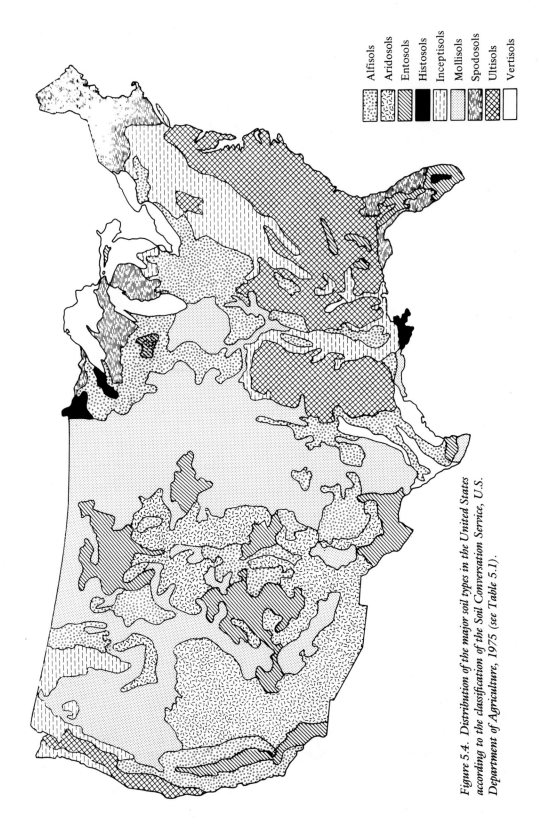

Figure 5.4. Distribution of the major soil types in the United States according to the classification of the Soil Conversation Service, U.S. Department of Agriculture, 1975 (see Table 5.1).

Soils, dusts, and muds

Laterites

Laterite is a special kind of oxisol that occurs in many tropical and semitropical regions of the world (Plate 31). The soil order of oxisols is characterized by an A horizon that has undergone extreme chemical weathering above a thick B horizon. The A layer of a laterite contains little humus because organic matter tends to decompose very rapidly in tropical climates. Both the A and B layers are rich in goethite and other hydrous ferric iron minerals.

If a laterite is removed from its environment and allowed to dry, it hardens to a bricklike material that cannot return to a soft, soil-like state. Indeed, the very name laterite is derived from the Latin word *later*, meaning brick or tile, and there is good evidence that laterite bricks have been used for buildings in tropical areas for millennia. More recently, the famous temples of Angkor Wat, in Cambodia, were constructed from laterite.

Lateritic soils, though useful for building purposes, are difficult, if not impossible, to cultivate. When tropical forests are removed and lateritic soils are plowed, the ground itself frequently hardens to a bricklike consistency and thus becomes unworkable.

Caliche

Caliche is the name given to the solid, almost impervious layer, composed largely of calcite, that forms at the top of the C soil horizon in many aridisols. Caliche is formed when one of the released elements, calcium, combines with carbon dioxide from the atmosphere to form calcite (Fig. 5.5). The calcite layer accumulates at the level to which groundwater rises. This happens because in arid and semiarid regions, water bringing dissolved salts up from below evaporates and deposits its chemical load when it gets near the surface.

MOVEMENT OF THE REGOLITH

The regolith, including soil, does not remain fixed and stationary over any geologically long period of time; forces continually move it about and redistribute it. During those processes, new rock is exposed to weathering and erosion, and consequently there is a more or less continuous cycle of uplift, weathering, removal, and dispersal of regolith. The chief dispersal agents are gravity, moving water, wind, and glaciers.

Gems, granites, and gravels

Figure 5.5. Caliche cementing gravel on the northern edge of the Namib Desert, Namibia. Caliche develops in dry climates through evaporation of groundwater and the consequent precipitation of dissolved calcium carbonates. (Photograph by B. J. Skinner)

Downslope movement

No large part of the Earth's surface is exactly horizontal. Nearly all plains slope in one or more directions, and hills and valleys are common features on all of the Earth's continents and most of its islands. Wherever there are slopes – even of a few degrees – gravity causes the regolith to creep or slip downslope. Indeed, a greater volume of regolith is moved as a consequence of gravity-driven processes than by any other means.

Movement by water

At the base of a slope, wherever a stream flows, the regolith that has slipped downslope is transported away by running water. In many cases the materials are subsequently transported by master streams into seas or oceans, where they are transported farther by water currents before being eventually deposited as sediments.

SANDS, GRAVELS, AND MUDS. Water frequently transports sands, gravels, and muds and deposits them as sediment. The slower the velocity of the water, the smaller the particles it can transport. Thus, as water velocity decreases, coarse particles can no longer be transported and thus are deposited, while finer particles continue to be carried farther downstream. In this manner, flowing water serves as a sorting and separating system, and thus we tend to find a sediment made up of coarse fragments

Soils, dusts, and muds

Figure 5.6. Clastic sediment. Many beaches usually covered by sand and gravel are covered by cobbles, with or without pebbles, after storms; the storm waves winnow out and carry the sand and sometimes also the pebbles offshore. This beach is along Lake Michigan in Charlevoix County, Michigan. (Photograph by R. E. Hampton)

in one place and sediment made up of fine particles in another place. It matters not whether the flowing water is a small stream, a large river, or a current in a lake or a sea or an ocean; the same principle applies.

As previously noted, fragments in the regolith are called clasts. The names given to clasts of diverse sizes are tabulated in Appendix 3 (Table A3.3). Note that the terms used for individual clasts differ from those used for aggregates.

Individual clasts may be either rock fragments or single mineral grains. The coarser a clast, the more likely it is to be a rock fragment. Among the many interesting places where one can examine a wide range of clastic sediments are the sand and gravel banks along lakeshores and seashores (Fig. 5.6). In most places the sands consist predominantly of quartz grains; the finer-grain-size fragments, if any are present, are largely clay minerals; the pebbles, cobbles, and boulders are nearly all tough igneous and metamorphic rocks. This is so because the continual pounding during water transport – whether by stream, current, or wave action – breaks up the softer and more easily broken rocks and the easily cleaved and

Gems, granites, and gravels

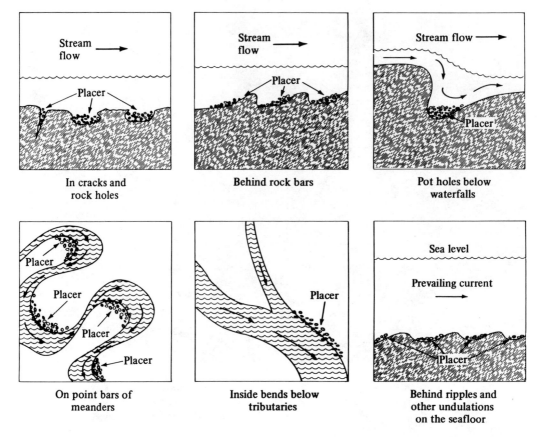

Figure 5.7. Placers are accumulations of dense, chemically and physically resistant minerals. They occur where a barrier to flow allows water to carry away lightweight particles while concentrating more dense particles behind the barrier.

fractured minerals. Only tough rocks persist as cobbles and pebbles, and only physically resistant minerals, such as quartz, remain as sand grains.

PLACERS. As mentioned in Chapter 4, the capacity of flowing water to transport clasts is not solely a function of grain size; it is also a function of the density of the clasts. The more dense a particle, the greater the velocity required to make it move. Therefore, flowing water will effect a separation of clasts on the basis of density as well as on the basis of size. The results of this process can be seen in action along many shorelines, especially during storms. As storm waves wash up onto the beach, the advancing wave comes in at a relatively high velocity, whereas its returning swash moves with a relatively low velocity. Thus, the returning water can transport only the relatively lighter and finer grains. As a result, a concentration of dark, fairly heavy grains of magnetite, garnet, and other dense minerals is left along the line of maximum wave advance.

Soils, dusts, and muds

Figure 5.8. The quest for gold has led to exploration, prospecting, and rushes for centuries. These turn-of-the-century prospectors were panning for gold in placer deposits in Alaska, a region famous for several "finds." (Courtesy Department of Geology and Geophysics, Yale University)

An accumulation of dense minerals formed as a consequence of stream flow or a lake or ocean current is called a *placer* (Fig. 5.7). Placers typically occur behind riffles in streams and in other places where water velocities are reduced. Some of the minerals concentrated in placers are valuable. Gold, for example, is dense as well as highly resistant to both chemical and mechanical weathering, and thus grains of gold often become concentrated in placers. In fact, more than half the gold that has been mined through all human history has come from placers (Fig. 5.8). Other minerals that are recovered from placers include platinum (especially in the USSR, near the Ural Mountains), diamonds (especially in Namibia, South Africa, the Central African Republic, and Zaire), rubies and sapphires (in Sri Lanka and Burma), and zircons (in Australia).

Movement by wind

The term *dust* refers to wind-transported regolith. Air, being less dense than water, has a much lower capacity for transporting clasts than running water does. Indeed, wind velocities seldom are high enough to transport coarse clasts like cobbles and boulders. In general, only dust grains, which are sand-sized or smaller, are moved.

Wind moves clasts in two ways. Material moved in suspension is called the *suspended load*. Material moved by bouncing and rolling is called the *bed load*. Sand dunes and sand ripples are local deposits made up predominantly of wind-deposited, bed-load clasts (Fig. 5.9).

Normally, a suspended load soon settles out; however, when winds are strong and flowing rather persistently, enormous dust storms can result. A well-documented, great dust storm started in western Kansas and eastern Colorado during the early morning hours of March 20, 1935. Twenty-seven hours later, by the middle of the day of March 21, the storm had reached New York and New England, about 2,400 km (~1,500 mi) away. So much of the dust remained suspended that twilight conditions existed in New York City at noon. The winds carrying the dust had to have averaged nearly 90 km/hour (~55 mph).

LOESS. A sediment made up entirely of windblown, suspended-load dust is called loess (\lə(r)s\, from the German word *Löss*). Loess is abundant here and there within the corn-growing areas of Illinois, Iowa, eastern Nebraska, and Missouri. That loess was deposited by winds that swept down off the great ice sheet that covered the northern half of North America at the height of the last ice age. The winds picked up a suspended load of fine particles from the glacial debris that was dumped by the ice near the edge of the ice sheet. As the winds flowed southward and decreased in velocity, they dropped that load as blankets of fine dust. Today, the soils developed on loess provide some of the richest agricultural areas in the region.

The world's most extensive region of loess deposits is in central China, where approximately 800,000 km^2 (~320,000 mi^2), are covered by a wind-deposited blanket of regolith that in places is said to be up to 300 m (~1,000 ft) thick (Fig. 5.10). Loess, however, is readily susceptible to erosion. Consequently, it is often transported and redeposited. The yellow color of the great Hwang Ho (Yellow River) in China is due to its load of loess, which has been lost to erosion because of a combination of natural conditions and poor farming practices in some of the loess-blanketed country within the river's basin.

Movement by ice

People who live in Europe north of the Alps, and in North America north of a line trending roughly from about New York City to Portland, Oregon, are familiar with a regolith of loose debris – ground-up rock

Soils, dusts, and muds

Figure 5.9. Windblown sand is common in deserts and along several low-lying coastal areas of large bodies of water. These dunes are along the Atlantic coast of eastern North Carolina. (Photograph by J. Kanode, courtesy of Norfolk & Western Railway)

Figure 5.10. A thick layer of loess covers underlying bedrock in the Yenan region, northern China. Landslide control and reduction of erosion are being attempted through building of terraces on which trees are planted. (Photograph by B. J. Skinner)

and boulders of all sizes – that was deposited by ice. Today it is common knowledge that only 25,000 years ago great ice sheets covered huge portions of the globe. That fact, however, has been recognized only rather recently.

Gems, granites, and gravels

Figure 5.11. Glacial till near Mt. Cook, South Island, New Zealand. Till consists of unsorted debris ranging in size from clay particles to boulders. (Photograph by R. V. Dietrich)

Throughout the 18th Century and during the early years of the 19th Century, geologists were puzzled by the blanket of unsorted clastic sediment that covered much of the ground in northern latitudes. That sediment includes, among other things, erratically distributed huge boulders that appeared to have been moved many miles. For a while, geologists accepted, albeit reluctantly, the idea that the boulders and sediment had been deposited by the Noachian flood. Such an explanation, however, was unsatisfactory so far as accounting for several other features, such as the smooth, polished surfaces of rocks that occur beneath the glacially deposited sediment. Therefore, by the 1820s, the deluge explanation was being questioned.

GLACIAL DRIFT. Charles Lyell, a famous English geologist, made one of the first alternative suggestions – that the sediment, the erratic boulders, and the polished rock surfaces were all the work of icebergs that had drifted in the floodwaters. To this day, we call that blanket of debris *drift*.

It remained, however, for the Swiss geologist Louis Agassiz to make the next important advance; he suggested that the drift was deposited by terrestrial ice rather than by icebergs. He also suggested that Greenland and Antarctica are even now covered by shrunken remnants of what were once much larger ice sheets. We now know that Agassiz's suggestions were correct. We also use the term *glacial drift* to include all debris whose deposition was largely dependent on glacial activities.

GLACIAL TILL.   As an ice sheet expands, and ice flows slowly outward from its center, the ice picks up regolith, including large boulders, and carries those fragments, no matter what their size. Because ice is so viscous, the clastic sediment it carries does not become sorted by size. When the ice melts, the entrained, unsorted debris is merely dumped in situ; consequently, the resulting unsorted deposits commonly consist of individual clastic grains that range from clay-sized particles up to boulders, some as large as houses. This material is called *glacial till* (Fig. 5.11).

Both thin sheets and huge piles of glacial till are called *moraines*. Well-known examples of moraines and till cover much of Long Island and Martha's Vineyard and many of the hilly areas in the upper midwestern United States.

GLACIAL ERRATICS.   The large boulders in till are usually called glacial *erratics*. A few erratics have become rather famous, either because of their size or because of the place they occur (Fig. 5.12). The brothers Grimm recorded legends about several of these erratics in central Europe.

STRATIFIED DRIFT.   Examination of nearly all glaciated regions reveals yet another kind of glacial drift besides till. This sediment constitutes the sporadic pockets of water-sorted sands and gravels that occur along with the unsorted till in glaciated terrains. These sorted sediments were deposited by flowing meltwater and thus have the same origin as other water-sorted sediments. They are, however, different: Their layers of sediment (i.e., their strata) occur within, upon, or marginal to glacial debris, and both the fragments and the water that carried those sediments came directly from the glacier. Consequently, these sediments are called *ice-contact stratified drift*, or merely *stratified drift* (Fig. 5.13). Many of the commercial sand and sand-and-gravel deposits in North America, Europe, and Asia are stratified drift.

Gems, granites, and gravels

Figure 5.12. "Big Rock" on Beaver Island, in northern Lake Michigan, is a boulder made up of an intrusive igneous rock that was glacially transported from north of Lake Superior, across what is now Lake Superior, and then across the upper peninsula of Michigan and northern Lake Michigan, to central Beaver Island, during the "Great Ice Age." Much of this boulder may be below the surface – that is, it is probably much larger than it appears to be. (Photograph by W. E. Moore)

Figure 5.13. Stratified drift deposit near Brier Hill, St. Lawrence County, New York. Concentrations of stratified drift like these sand and gravel beds were deposited at many places by streams flowing on, within, or adjacent to glacial ice during the "Great Ice Age." Several of the deposits, including this one, have furnished large tonnages of sand and gravel, especially for use as the aggregate in concrete. (Photograph by R. V. Dietrich)

FURTHER READING

Birkeland, P. W., 1984, *Soils and Geomorphology*. Oxford University Press, 372p.
A well-written, interesting treatment of soils and how their development is controlled by geomorphology – that is, the underlying geology and the lay of the land.

Bridges, E. M., 1978, *World Soils*. Cambridge University Press, 128p.
A global view of soil types and their distribution.

Buol, S. W., Hole, F. D., and McCracken, R. J., 1980, *Soil Genesis and Classification* (2nd edition). Iowa State University Press, Ames, 360p.
A professional treatise dealing with essentially all aspects of the chemistry and physics of soil development. A discussion of the soil classification system currently used in the United States is included.

Dietrich, R. V., 1989, *Stones: Their Collection, Identification, and Uses* (2nd edition). Geoscience Press, Prescott, Arizona, 191p.
This inexpensive paperback includes descriptions of loose mineral and rock fragments, where they occur, how they were formed, and how they are used.

Flint, R. F., 1971, *Glacial and Quaternary Geology*. Wiley, New York, 892p.
An authoritative volume covering essentially all aspects of glaciation during the recent Great Ice Age.

Friedman, G. M., and Sanders, J. E., 1978, *Principles of Sedimentology*. Wiley, New York, 792p.
Describes diverse sedimentary processes and deposits.

Press, F., and Siever, R., 1986, *Earth* (4th edition). W. H. Freeman, San Francisco, 656p.
A comprehensive elementary textbook of geology.

Revelle, R., 1984, The world supply of agricultural land, pp. 184–201 in Simon, J. L., and Kahn, H. (editors), *The Resourceful Earth*. Blackwell Scientific, Oxford, England, 585p.
An interesting essay on the problems and challenges of soil usage facing us as we farm ever more heavily and clear more and more land in order to do so.

Skinner, B. J., and Porter, S. C., 1987, *Physical Geology*. Wiley, New York, 750p.
An up-to-date, extremely well illustrated basic textbook of geology.

6
Ores and ore minerals

Our nomadic ancestors knew how to shape rocks, how to fashion implements from wood and bone, and how to make pottery by baking clay on open fires. Some of the utensils dug up by the archaeologists who study Stone Age people are rather intricate and bespeak the highly developed skills of their makers. Such skills indicate that Stone Age societies must have been complex organizations in which stoneworkers, boneworkers, woodworkers, and other specialists all played roles.

One day, some unrecorded ancestor picked up a stone of metallic copper – such stones, though not common, do exist – and discovered that it was malleable, that it could be fashioned into more intricate shapes than was possible with wood, bone, or brittle rock. The Stone Age came to an end.

We can only imagine the scene when metals were first discovered. We have no knowledge as to exactly where, when, or by whom the discovery was made. We cannot even be sure that copper was the metal first discovered and used; perhaps gold or silver was first. Archaeological data are not clear on these points. We do know, however, that the systematic use of metals started more than six thousand years ago, because by 4000 B.C. the Chaldeans, who lived in a province of Babylonia near the head of the Persian Gulf, had already established a metal industry (Plate 32). Indeed, the Chaldeans' metalworking skills, which included copper, gold, silver, lead, and tin, gave them a highly esteemed reputation as metalworkers among the ancient peoples of the Middle East.

Perhaps the first use of metal occurred in the Middle East; perhaps it was elsewhere – possibly in one of the more ancient civilizations of southeastern Asia. Regardless of where the initial discovery was made, information about metals and metalworking was soon known here and there the world over. In fact, by the time of the Chaldeans, many groups of people knew how to extract certain metals from their ores, how to mine those ores, how to cast and work metals, and even how to mix

Ores and ore minerals

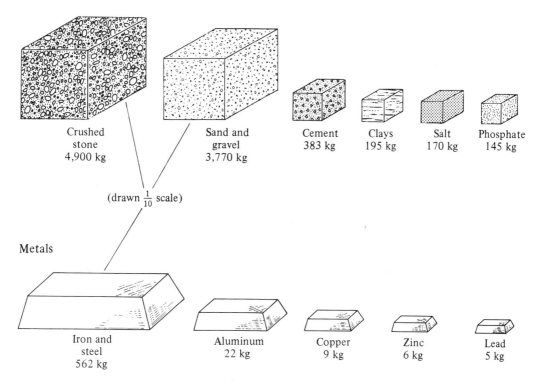

Figure 6.1. The amount of material used by every man, woman, and child in the United States in 1988. The range of materials mined today is very large, and the total amount is enormous. Approximately 17 tons of new mineral materials are now required annually by each U.S. citizen.

metals in order to make alloys, such as bronze (copper plus tin), that were tougher than the pure metals from which they were made.

KINDS OF RESOURCES

The discovery of metals changed human society forever. From the hypothesized earliest discovery of stones consisting of pure native copper to the present day, when we use a metal, such as hafnium, that was discovered only during the 20th Century, humankind has found uses for more and more of the *metallic resources* that can be taken from the Earth (Fig. 6.1).

Not all of the materials we use, however, are metals. We also mine raw minerals and rocks that are used in a multitude of other materials, including such things as cement, plaster of Paris, fertilizers, and even the

WHAT MAKES AN ORE?

All ores are mineral deposits, but not all mineral deposits are ores. The term **mineral deposit** is a scientific term; it means any volume of rock containing an enrichment of one or more minerals. **Ore deposit** is an economic term; it designates a mineral deposit that can be worked at a profit.

Whether or not a given mineral deposit is an ore depends on many things. Most obviously the size and richness of the deposit are very important. But so, too, are such things as geographical location and the cost of labor. Consider **geography**: The costs of transporting equipment and materials to a mine and mineral products away from a mine to the marketplace must always be considered. Even a very rich deposit – for example, one high in the remote Andes or in the central Sahara – might not be an ore, because transportation costs might be too high. Also, **labor costs** differ from country to country by factors of ten or even more; thus, a deposit that could be mined in, say, Zaire, where labor costs are low, might not be minable at a profit in France, where labor costs are high.

Other major factors that determine whether a given deposit is or is not an ore are the political and economic stability of the country where the deposit occurs and the market price of the metal produced. Political and economic stability must be considered when investors risk the hundreds of millions of dollars needed to open a mine. As for market price, consider copper as an example. In recent years, the price of copper has fluctuated between 60 and 165 cents (U.S.) per pound. When its price has been high, even lean, difficult-to-work deposits have been ores; on the other hand, when the price has been low, only the richest and least costly mines have been profitable to operate. Consequently, a deposit can be an ore one day and lose that status the next, just because of some change in one or more nongeological factors.

Yet another factor, technology, can make or break an ore. For example, for many years it was possible to find iron ores that were so rich they could be fed directly into blast furnaces without prior treatment. By the end of World War II, however, that kind of rich ore was nearing exhaustion within the United States. Technology saved the day. A method was developed to crush lean siliceous ores called taconites, to separate inexpensively the constituent iron minerals from the valueless quartz of these taconites, and to compress the separated iron mineral particles into compact pellets that make ideal blast-furnace feed. Indeed, taconite pellets are such good feed that the cost of smelting has been reduced, and pellets are now the standard for the industry. Furthermore, billions of tons of taconite that once were considered waste now qualify as ore!

---

mineral fillers in the paper on which these words are printed. We refer to materials mined for purposes other than the recovery of metals as *nonmetallic resources*.

It is, however, the metallic resources that have really changed our lives and are the very basis of our society. Sometimes we tend to think that *energy resources,* such as coal and oil and natural gas, are the keys to modern society. But think about it: What would we do with coal or oil if we lacked metals? Is it not true that in order to use energy resources we need machines, and that to build machines we need metals? Also, do we not need machines to use most of the nonmetallic resources? Indeed, the uses of all *mineral resources* – metallic, nonmetallic, and energy – are interrelated, but the key ingredients in the system are metals. We have much for which to thank our inquiring ancestors who discovered most of the metals we use today.

ORE MINERALS

The reason that mines exist is straightforward. The places where mineral resources are found are small localized bodies of rock that are enriched in one or more given minerals. Why a local enrichment of a given mineral resource occurs in one place but not in another is the result of complex sequences of geological events. The sequence and complexity of resource-forming processes differ from one enrichment to another – indeed, each is unique. Nonetheless, it is possible to group enrichments into distinct families based on their compositions, the processes by which they were formed, or other characteristics. The most important characteristic, however, is readily recognized: It is the kind of mineral or minerals present. Hence, we tend to use the term *deposit,* along with an appropriate adjective or two that indicate the chief product and/or, for example, the way the deposit is thought to have originated.

People who work in the mineral resource business refer to mineral deposits that are large enough and rich enough to be worked at a profit as *ore deposits*. An ore deposit can be described simply as a localized body of rock that contains a mixture of economically valuable *ore minerals* and valueless *gangue* (pronounced \găŋ\) minerals (Fig. 6.2).

Once an ore deposit has been discovered, the task is to dig out the ore as efficiently as possible (Fig. 6.3) and then separate the ore minerals from each other and from the gangue. The miner's goal is a pure concentrate of each ore mineral. Naturally, then, the most desirable ore deposits are those in which the ratio of ore minerals to gangue is highest and from which the ore minerals can be separated cleanly.

Once a concentration of ore minerals has been effected and the gangue discarded, the desired metal or metals usually have to be recovered by smelting the concentrates. Smelting, a chemical process whereby the desired metal is separated from its containing mineral, is frequently the

Figure 6.2. Copper ore from Butte, Montana. The host is an igneous rock, now highly altered by the hydrothermal solutions that carried in chalcopyrite ($CuFeS_2$), the chief ore mineral, which is seen here filling a network of thin dark veins. (Photograph by B. J. Skinner)

most expensive and difficult step in the production of a metal (Plate 33). Some minerals, of course, are easier to smelt than others; those that are richest in the desired metal and are easiest to smelt are called ore minerals. Most of the familiar metals, such as iron, aluminum, copper, lead, zinc, nickel, gold, and silver, are present in at least trace amounts in tens or even hundreds of different minerals. Nearly all metals, however, have only four or five important ore minerals, and the less familiar metals, such as zirconium and tantalum, generally have only a single ore mineral.

The most important ore minerals for metals are listed in Table 6.1. Notice that many of them are sulfides (e.g., chalcopyrite, $CuFeS_2$, the main ore mineral of copper), oxides (e.g., cassiterite, SnO_2, the main ore mineral of tin), or native metals (e.g., gold, Au, and platinum, Pt). Also notice that very few ore minerals are silicates, even though the Earth's

Ores and ore minerals

Figure 6.3. Santa Rita, New Mexico, a modern, open-cut copper pit. Several metallic and nonmetallic ores are recovered from open pits rather than from underground mines, which are widely believed to yield most mineral resources. (Photograph by B. J. Skinner)

crust is composed largely – at least 98 percent – of silicate minerals. The reason for this is simple: Silicate minerals are exceedingly difficult to smelt and thus are used only when no alternative sources exist. An example of a silicate ore mineral is zircon ($ZrSiO_4$), the only ore mineral of zirconium.

METALLIC ORE DEPOSITS

When prospectors of yore talked about finding a bonanza, they were referring to the discovery of a rich ore deposit from which metals could be produced. They were not referring to a deposit of salt, sulfur, or some other nonmetallic resource. For many prospectors, gold or silver was their only goal. Ironically, far more wealth has come from the mining of the less glamorous metals than has ever accrued from gold and silver. Consequently, copper and other metals eventually also became the targets of some prospectors' searches. All of the metallic ore deposits thus far discovered were formed in one of four different ways.

Deposits formed by hydrothermal solutions

We live on an unusual planet – there is no similar body in our solar system. The main reason the Earth is unique is that its atmosphere, its size, and its distance from the sun combine to produce the narrow range

Table 6.1. *Principal metals and their ore minerals*

Metal	Ore minerals	Kinds of deposits	Major producing countries
Aluminum	Gibbsite, $Al(OH)_3$ Diaspore, $AlO(OH)$	Residual	Jamaica, Australia, Guinea
Antimony	Stibnite, Sb_2S_3 Tetrahedrite, $Cu_{12}Sb_4S_{13}$	Hydrothermal	China, USSR, South Africa
Beryllium	Beryl, $Be_3Al_2Si_6O_{18}$ Bertrandite, $Be_4Si_2O_7(OH)$	Pegmatite, hydrothermal	USA, USSR, Brazil
Bismuth	Bismuthinite, Bi_2S_3	Hydrothermal	Australia, Peru
Cesium	Pollucite, $Cs_2Al_4Si_4O_{12}\cdot H_2O$	Pegmatite	Canada, Zimbabwe
Chromium	Chromite $(Fe, Mg)Cr_2O_4$	Magmatic segregation	South Africa, USSR
Cobalt	Linnaeite, Co_3S_4	Hydrothermal	Zaire, Zambia
Copper	Chalcocite, Cu_2S Chalcopyrite, $CuFeS_2$ Bornite, Cu_5FeS_4 Enargite, Cu_3AsS_4	Hydrothermal	Chile, USA, USSR, Canada, Zaire
Gold	Native gold, Au	Placer, hydrothermal	South Africa, USSR, Canada, USA, Australia
Iron	Goethite, $FeO(OH)$ Hematite, Fe_2O_3 Magnetite, Fe_3O_4 Siderite, $FeCO_3$	Chemical sediment, residual	USSR, Brazil, Australia, China
Lead	Galena, PbS	Hydrothermal	Australia, USSR, USA
Lithium	Spodumene, $LiAlSi_2O_6$ Eucryptite, $LiAlSiO_4$	Pegmatite	USSR, China, Brazil
Magnesium	Dolomite, $CaMg(CO_3)_2$ Magnesite, $MgCO_3$	Chemical sediment, residual	USA, USSR, Norway
Manganese	Pyrolusite, MnO_2 Wad, "hard Mn oxides"	Chemical sediment	USSR, South Africa, Brazil
Mercury	Cinnabar, HgS Metacinnabar, HgS	Hydrothermal	USSR, Spain, Algeria
Molybdenum	Molybdenite, MoS_2	Hydrothermal	USA, Chile, USSR
Nickel	Pentlandite, $(Fe, Ni)_9S_8$ Garnierite (mixture of hydrous Ni silicate minerals)	Magmatic segregation, residual	USSR, Canada, New Caledonia, Australia
Niobium	See Tantalum		
Platinum group metals[a]	Native platinum, Pt Sperrylite, $PtAs_2$	Magmatic segregation	South Africa, USSR
Rare earths[b]	Monazite, $(Ce, La, Th)PO_4$	Pegmatite, carbonatite	Australia, Brazil, USA

Table 6.1. (cont.)

Metal	Ore minerals	Kinds of deposits	Major producing countries
	Bastnasite, (Ce, La)(CO$_3$)F Cerite, Ce$_9$MgSi$_7$O$_{28}$ (Cerium, Ce, and lanthanum, La, here represent all of the rare earths; all are present to some extent in monazite.)		
Silver	Native silver, Ag Argentite, Ag$_2$S Miagyrite, AgSbS$_2$	Hydrothermal	Mexico, Peru, USSR, USA, Canada
Tantalum–niobium	Columbite–tantalite, (Fe, Mn)(Nb, Ta)$_2$O$_6$ Pyrochlore, (Ca, Na)$_2$Nb$_2$O$_6$(OH, F)	Hydrothermal, pegmatite	Brazil, Canada
Thorium	Zircon, (Zr, Th)SiO$_4$ Monazite, (Ce, La, Th)PO$_4$	Placer	Australia, Brazil, Malaysia
Tin	Cassiterite, SnO$_2$	Hydrothermal, placer	Malaysia, USSR, Brazil, Thailand
Titanium	Rutile, TiO$_2$ Ilmenite, FeTiO$_3$	Placer, magmatic segregation	Australia, Norway, USSR
Tungsten	Wolframite, (Fe, Mn)WO$_4$ Scheelite, CaWO$_4$	Pegmatite, hydrothermal	China, USSR, South Korea
Uranium	Uraninite, UO$_2$ Brannerite, UTi$_2$O$_6$ Carnotite, K$_2$(UO$_2$)$_2$V$_2$O$_8$·3H$_2$O	Hydrothermal	USA, Namibia, Australia, Canada
Vanadium	Vanadiferous magnetite, (Fe, V)$_3$O$_4$	Magmatic segregation	South Africa, USSR, China
Zinc	Sphalerite, ZnS	Hydrothermal	Canada, Australia, Mexico
Zirconium and hafnium	Zircon, (Zr, Hf)SiO$_4$	Placer	Australia, South Africa

[a] The platinum group metals are platinum, palladium, osmium, iridium, rhodium, and ruthenium. They have similar properties and are found together in the same minerals.
[b] The rare-earth metals are the fourteen chemical elements, atomic numbers 57 through 71 in the periodic table. They tend to occur together in the same ore minerals because they have very similar chemical properties. The term "rare" is a misnomer, because the rare-earth metals are more abundant than some of the familiar metals, such as lead. The rare-earth metals are cerium, Ce; dysprosium, Dy; erbium, Er; europium, Eu; gadolinium, Gd; holmium, Ho; lanthanum, La; lutetium, Lu; neodymium, Nd; praseodymium, Pr; promethium, Pm; samarium, Sm; terbium, Tb; thulium, Tm; ytterbium, Yb.

of temperatures and pressures required for water to remain at its surface in the liquid state. The water that falls as rain accumulates in the oceans, lakes, and streams and even permeates down from the surface to fill tiny openings within rocks of the Earth's crust. The presence of this magical fluid, both at and below the surface of the Earth, is the main reason that Earth has so many different kinds of ore deposits. Water not only facilitates but also is essential to several of the chemical reactions that are responsible for the formation of many ore deposits. In fact, the absence of water is the reason that such deposits cannot form on our dry moon, on Venus, with its intensely hot atmosphere, or probably even on Mars, although Mars may once have had running water on its surface.

We now know that water is present in Earth rocks down to great depths. Water was found in the tiny openings between mineral grains and in every crack and fracture in the rock brought up from the deepest hole yet drilled into the Earth – a hole on the Kola Peninsula in the USSR that was 13.2 km (~8 mi) deep. We surmise that at least tiny amounts of water are present at even greater depths.

The water in rocks is, of course, not pure. This is so because temperature increases with depth, and the heated water extracts soluble constituents from the enclosing rocks. Indeed, most of the water within the Earth's crust is a solution of salts such as $NaCl$, $CaCl_2$, and $MgSO_4$; in fact, much of the water in rocks eventually becomes a brine because at least small amounts of soluble material are present in most rocks. In any case, even water with only a small amount of a dissolved salt (e.g., $NaCl$), is a potent solvent for metals, and therein lies the key to this most important family of metallic ore deposits.

As the Earth's crust is moved around – raised, lowered, intruded by hot masses of magma, and otherwise disturbed – deeply penetrating water solutions flow from one place to another. Most of the water flow is exceedingly slow. However, in some places – such as around volcanic vents, where eruptions shatter rocks and increase their permeability, and along great fractures, such as those that rift the ocean floor – there are well-defined flow channels, and hot, deeply buried fluids can rise up rather rapidly through these channelways, and as they do they react chemically with the rocks that line the channel walls. Eventually, some of the hot rising solutions reach the surface and form hot springs, like those at Yellowstone National Park (Fig. 6.4) or Rotorua in New Zealand. Beneath the sea, similar hot springs form the curious "smokers," recently discovered by deep-diving submarines at water depths of about 2.5 km (~8,250 ft) along the East Pacific Rise (Plate 34). All of these hot solutions, whether moving underground or issuing forth from springs on land or beneath the seas, are called *hydrothermal solutions*.

The dissolved materials that hydrothermal solutions carry are depos-

Figure 6.4. Grotto Geyser, Yellowstone National Park, Montana. The geyser, here seen erupting steam, is the top of a large, active, hot-spring system. Hot waters such as these are responsible for the deposition of many minerals at the surface, underground, and even under seas and oceans. (Photograph by B. J. Skinner)

ited as the solutions cool and/or react with the rocks that line the flow channels. If the flow and deposition occur in an open fracture, the result is a *vein* made up of one or more precipitated minerals; if the flow is through a mass of closely shattered rock and deposition occurs, a *disseminated ore deposit* is formed.

Evidence proving that hydrothermal solutions react with rocks that line the flow channels can be found in zones of leached and chemically altered rock (Fig. 6.5). Such zones of hydrothermal alteration tend to be rich in clay minerals because the chemistry of hydrothermal alteration is similar to that caused by weathering. Some bodies of rock become so pervasively altered to masses of clay that the alteration zone itself becomes a nonmetallic resource – for example, the famous china-clay mines of Cornwall, England, had this origin.

Hydrothermal ore deposits comprise the most common kinds of metallic ores. Nearly all of our copper, lead, zinc, gold, silver, mercury, molybdenum, tin, tungsten, uranium, cobalt, germanium, gallium, and a host of other less important metals have come from these deposits.

Deposits formed in igneous rocks

Magma is molten rock material. As already mentioned, volcanoes, great bodies of granite, and many other distinctive geological features form from magma. Magma is generated when lower portions of the Earth's

Figure 6.5. Alteration produced by hydrothermal solutions, Gaspé Copper Mine, Québec. The host rock is an impure limestone. Solutions moving through the fractures produced a halo of alteration. (Photograph courtesy of John Allcock)

Figure 6.6. Layers of chromite (black), the only important chromium ore mineral, formed by magmatic segregation during crystallization of a layered, igneous intrusion. White layers are a plagioclase feldspar-rich rock called anorthosite. The location is the Dwars River, Transvaal, South Africa. (Photograph courtesy of Department of Geology and Geophysics, Yale University)

crust or the underlying upper mantle reach temperatures sufficiently high for melting to occur. Because the materials that melt are ordinary rocks, most magmas are not notably enriched in ore-forming constituents. But magmas can become so enriched, and where they do, ore deposits can form as those magmas cool and crystallize. These deposits form in two ways.

Each magma, no matter how it originates, is a complex liquid. Thus, it does not solidify at a single temperature the way water freezes to ice at 0°C (32°F). Instead, a magma freezes over a wide range of temperatures as first one mineral and then another crystallizes. If one of the early minerals to crystallize is an ore mineral – for example, chromite [$(Fe, Mg)Cr_2O_4$], the main source of chromium (Fig. 6.6) – and that mineral is more dense than the remaining magma, grains of the mineral will sink to the bottom of the magma body and accumulate there. Thus, when the entire body of magma has crystallized and cooled, it will include a layer in which the mineral has been concentrated. This concentrating process is called *magmatic segregation*. It is the principal way in which ore deposits of chromium, vanadium, titanium, nickel, platinum, palladium, osmium, iridium, and ruthenium are formed.

A late-stage effect can lead to another kind of concentration of valuable minerals: Crystallization of the bulk of a typical magma results in the formation of more or less valueless minerals, such as quartz and feldspar. As those minerals crystallize, however, a so-called *residual magma* is formed. In most cases, such an end-stage residual magma comprises only 1 percent or less by volume of the original magma. It is, however, an important portion, because relatively rare metals such as beryllium, tantalum, and niobium, which were present in only trace amounts in the parent magma, can be concentrated in these fluids. In some cases, when the end-stage fluid crystallizes, it forms small bodies of igneous rock of very coarse grain size called *pegmatite* (Fig. 6.7), and crystals of minerals containing one or more of the rare chemical elements often occur in these rocks. This process of concentration is called *magmatic fractionation*.

Some pegmatite masses are so enriched in a given ore mineral that the entire body can be mined; more commonly, pegmatites consist of more than one mineral, but are so coarse-grained that they are mined selectively, mineral grain by mineral grain. The principal metals recovered from pegmatites are beryllium, niobium, tantalum, lithium, and cesium. Pegmatites are also the sources of many gemstones and have yielded some of the largest and most striking mineral specimens that are the central showpieces in many museums.

Deposits formed by sedimentation

Material dissolved in seawater or in lakes can be precipitated in either of two ways:

1. by evaporation of water in a restricted arm of a sea or within a closed lake to form a chemical sediment, or
2. in response to some kind of biochemical activity to form a biochemical sediment.

Gems, granites, and gravels

Figure 6.7. Many of the world's largest crystals have been formed, apparently by relatively slow growth, in pegmatite masses. Examples are the 50-ft-long spodumene ($LiAlSi_2O_6$) crystal at the Etta Mine, Keystone, South Dakota, and the highly valued, nearly 4-ft-long, cranberry-colored tourmaline [variety, elbaite ~$Na(Al, Li)_3Al_6B_3Si_6O_{27}(O, OH, F)_4$] crystal from the Itatiaia district, Minas Gerais, Brazil. The large, white plagioclase feldspar (albite) crystal shown is in a pegmatite near Spruce Pine, North Carolina. (Photograph by J. E. Callahan)

Many chemical and biochemical sediments form nonmetallic mineral resources – for example, rock salt ($NaCl$), gypsum ($CaSO_4 \cdot 2H_2O$), limestone ($CaCO_3$), and rock phosphate [$Ca_5(PO_4, CO_3)_3(F, OH)$]. Under some circumstances, however, minerals rich in iron or manganese are precipitated, and the end result can be a sediment rich in an iron ore mineral, such as goethite [$FeO(OH)$] or hematite (Fe_2O_3) (Plate 35), or a manganese ore mineral, such as pyrolusite (MnO_2) or todorokite [$(Mn, Ca, Mg)Mn_3O_7 \cdot H_2O$]. All such deposits are called *chemical sedimentary ores*.

Chemical sedimentary ores of both iron and manganese have formed many times during geological history. In fact, most of our modern supplies of both iron and manganese come from such deposits. Today, so far as we know, no large chemical sedimentary iron ore deposits are being formed. There are, however, modern chemical sedimentary manganese ore deposits – they are made up of curious black nodules of manganese oxide that are forming here and there on the floor of the deep ocean (Fig. 6.8).

Ores and ore minerals

Figure 6.8. Cross section of a manganese nodule from the floor of the Pacific Ocean. Note the concentric layers due to deposition of numerous manganese and ferromanganese hydroxide layers. (Photograph by W. Sacco)

Deposits formed by weathering and erosion

As discussed in Chapter 4, the processes of chemical weathering involve chemical alterations of rocks that take place when the rocks interact with rainwater or are penetrated by groundwater. Chemical elements such as potassium, sodium, calcium, and magnesium are rather easily removed by hydrolysis, and eventually even quartz can be dissolved. If the weather is warm, as in the tropics, and the process persists over any relatively long period of time, only the least soluble material may eventually remain as an enriched residue. If the residual material is an ore mineral, a mineral deposit results. Mineral deposits formed by this process are called *residual deposits*. The most abundant and commonly observed mineral resulting from such enrichment is the ferric iron mineral goethite [FeO(OH)], the principal mineral present in laterite (see Chapter 5). In some places, such as in western Africa, the enrichment of iron in laterite is so great that potential iron ore deposits have been formed. Although, to date, none of these laterites has been worked for iron on any large scale, it seems likely that residual ores of iron will be worked at some time in the future (Fig. 6.9).

Under certain circumstances – such as a tropical climate with a seasonal rainfall plus a flat or rolling topography and a light vegetation cover – even iron minerals can be removed in solution. Then, only the least soluble minerals remain in the residue. These are the aluminum hydroxide minerals, gibbsite [Al(OH)$_3$] and diaspore [AlO(OH)]. Either alone or

Gems, granites, and gravels

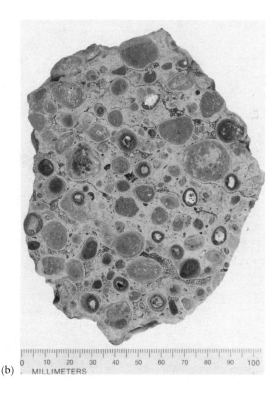

(a)   (b)

Figure 6.9. Nodular structures are common in many laterites. **(a)** Iron-rich laterite from Western Australia; specimen is 10 cm in length. **(b)** Aluminum-rich laterite (bauxite) from Arkansas. (Photographs courtesy of W. Sacco)

in a mixture, these two minerals constitute the chief ore of aluminum, *bauxite*. All bauxites occur as residual deposits.

All exposed rocks may be affected by weathering. When weathering alters preexisting hydrothermal and magmatic ore deposits, spectacular mineral displays may arise. The chemical weathering of an ore deposit is a process much like the alteration of a more common rock or rock material to form soil: At the surface there is a cap of insoluble minerals from which soluble materials have been leached; below the leached cap there is a zone where the leached materials become concentrated, the *zone of secondary enrichment*; below the zone of concentration there is the more or less unchanged original rock, the *primary ore*. Leached cappings tend to consist largely of insoluble masses of goethite and manganese oxides; these cappings are called *gossans*. Prospectors discovered long ago that prospecting beneath gossans was often rewarding. What can be discovered beneath a gossan, if one is lucky, is a mass of highly enriched ore minerals, in the zone of secondary enrichment.

The gossan, like the A zone of a soil, has been leached of most of its soluble constituents; the zone of secondary enrichment, like the B zone of a soil, is the place where much of the leached material is redeposited; the underlying primary ore is the unweathered precursor rock. Many gossans and zones of secondary enrichment contain spectacular mineral specimens. Malachite and azurite (the green and blue carbonates of copper, $Cu_2CO_3(OH)_2$ and $Cu_3(CO_3)(OH)_2$, respectively), cerussite ($PbCO_3$) (Plate 36), smithsonite ($ZnCO_3$) (Plate 37), and many other striking minerals occur in some gossans. Exceptional specimens of chalcocite (Cu_2S) and silver minerals, such as pyragyrite (Ag_3SbS_3), are found in some zones of secondary enrichment.

The final manner in which mineral concentration can occur as a result of weathering and erosion is in *placers*, which were mentioned in Chapter 5. In order for a placer to form, an ore mineral must be chemically resistant, so that it will remain little changed by chemical weathering, and dense, so that its transportation and deposition by flowing water or by lake or ocean currents will lead to its concentration. The principal metallic ore minerals that occur in placers are native gold, native platinum, zircon, and rutile and ilmenite (the chief ore minerals of titanium).

NONMETALLIC ORE DEPOSITS

Nonmetallic ore minerals are mined and used for their properties or their nonmetallic element content rather than for any metallic elements they may contain. An example of a nonmetallic ore mineral that is important because of its intrinsic properties is halite (NaCl, common salt); it is used for its taste, for preserving food, and for melting ice. An example of a nonmetallic ore mineral that is critical as raw feed for the chemical industry is apatite $[Ca_5(PO_4, CO_3)_3(F, OH)]$; it is the principal source of phosphatic fertilizers.

Nonmetallic ore minerals used in the chemical industry, and for making fertilizers, tend to be halides (i.e., compounds of the halogen elements: chlorine, bromine, iodine, and fluorine), carbonates, sulfates, borates, and phosphates. These minerals are listed in Table 6.2. The main ore minerals used in the glass and ceramic industries are quartz, feldspar, and clay minerals. Glass and the ceramic materials, brick and tile, are described in Chapter 7. Here, we only briefly discuss the major minerals of the chemical and fertilizer industries.

Table 6.2. *Principal nonmetallic ore minerals used in the chemical and fertilizer industries*

Chemical element of interest	Ore minerals	Kinds of deposits	Major producing countries
Barium	Barite, $BaSO_4$	Hydrothermal	China, USA, India, USSR
Boron	Borax, $Na_2B_4O_7 \cdot 10H_2O$ Colemanite, $Ca_2B_6O_{11} \cdot 5H_2O$ Ulexite, $NaCaB_5O_6(OH)_6 \cdot 5H_2O$	Evaporite (lake)	USA, Turkey, USSR
Bromine	Produced from brines	Evaporite (marine)	USA, Israel
Calcium	Calcite, $CaCO_3$	Chemical sediment	Essentially all countries
Fluorine	Fluorite, CaF_2	Hydrothermal	Mexico, Mongolia, China, USSR
Magnesium	Magnesite, $MgCO_3$ (also produced from brines)	Residual and evaporite (marine)	USA, China, North Korea, USSR
Phosphorus	Apatite, $Ca_5(PO_4, CO_3)_3(F, OH)$ (main mineral of rock phosphate)	Chemical sediment	USA, USSR, Morocco, China
Potassium	Sylvite, KCl Carnallite, $KMgCl_3 \cdot 6H_2O$	Evaporite (marine)	USSR, Canada, East Germany
Sodium			
Halide	Halite, NaCl	Evaporite (marine)	USA, USSR, China, Canada
Carbonate	Trona, $Na_3(CO_3)(HCO_3) \cdot 2H_2O$ Natron, $Na_2CO_3 \cdot 10H_2O$	Evaporite (lake)	USA, USSR, China
Sulfate	Thenardite, Na_2SO_4 Glauberite, $Na_2(SO_4)_2$	Evaporite (lake)	Mexico, Canada, USSR
Sulfur	Native sulfur, S Sour gas (e.g., H_2S)	Petroleum	USA, USSR, Canada, Poland

Ores and ore minerals

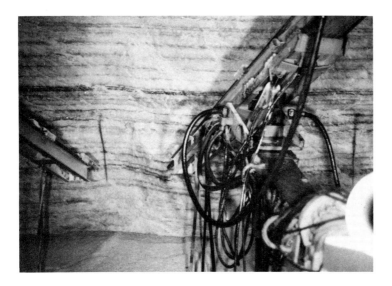

Figure 6.10. Evaporite deposit. Most of the world's salt is mined from deposits like this one in the Pugwash Mine, Nova Scotia, Canada. The machine is a horizontal drilling machine. (Photograph courtesy of Morton Salt Division, Morton International, Inc.)

Evaporite deposits

Substances that dissolve readily in water or weak acids are easier to treat chemically than insoluble substances that must be smelted. Most nonmetallic ore minerals are such soluble compounds.

Where might one expect to find soluble compounds in nature? In those great solvent ponds: the oceans and closed lakes. Soluble compounds that are dissolved during rock weathering processes are transported by stream waters and accumulated in the waters of oceans and lakes, and subsequently, when these restricted bodies of oceanic or lake water are subjected to evaporation, the soluble salts they hold can be released from solution and thus form extensive sedimentary layers. The layers made up of mineral salts formed through such evaporation are called *evaporite deposits* (Fig. 6.10).

Evaporite deposits are widespread and highly important. Most of the world's rock salt (the mineral halite, NaCl) is mined from marine evaporite beds. So, too, are the world's supplies of potassium salts for fertilizers (an important mineral is sylvite, KCl), of gypsum ($CaSO_4 \cdot 2H_2O$) for plaster, and so forth. Deposits formed as a result of the evaporation of ancient lake waters yield such minerals as trona [$Na_3(CO_3)(HCO_3) \cdot 2H_2O$] and thenardite ($Na_2SO_4$), both of which are sources of sodium (Na), which is used in many chemical processes, and of borax ($Na_2B_4O_7 \cdot 10H_2O$) and colemanite ($Ca_2B_6O_{11} \cdot 5H_2O$), both of which are sources of boron (B).

Biochemical deposits

Several kinds of important nonmetallic ore deposits form in response to the activities of living animals or plants. Rock phosphate deposits, for example, form where deep ocean waters well upward to the surface and in the process become so supersaturated that phosphate compounds are precipitated. The deep bottom waters of the ocean are phosphate-rich because dead animal and plant matter that falls to the sea floor decays and releases the phosphates contained in its cells to the seawater solution. The major phosphate-rich upwellings in today's oceans are along the west coast of South America and the west coast of Africa. In the past, however, there have been many areas of phosphate upwelling, and consequently phosphatic sediments are widespread geographically as well as geologically – that is, they occur sporadically in sedimentary rock sequences of many geological ages.

Another nonmetallic resource concentrated through biochemical agencies is sulfur. Certain bacteria live under anaerobic conditions, but even so must get oxygen from somewhere in order to survive. Some of them meet their needs by stripping oxygen away from sulfate (SO_4) compounds. We say that these bacteria have *reduced* the sulfate to native sulfur. Sulfate-reducing bacteria, like all bacteria, also need a source of food, which means a source of organic matter. One of the environments where these bacteria can find both sulfate compounds (usually gypsum, $CaSO_4 \cdot 2H_2O$) and organic matter (usually crude oil) is within certain evaporite deposits and in the long, fingerlike bodies of salt called *salt domes* that have risen upward from evaporite beds to pierce overlying oil-bearing strata. In such locations, whenever the conditions are right, these bacteria reduce the sulfate – in this case, the $CaSO_4 \cdot 2H_2O$ – to form masses of brilliant yellow native sulfur (Plate 38).

Hydrothermal deposits

Certain nonmetallic ore deposits form by deposition from hydrothermal solutions, which were described in a previous section of this chapter. There are two major examples: fluorite (CaF_2) deposits (Plate 39a) and barite ($BaSO_4$) deposits (Plate 39b).

Fluorite is used to make fluxes utilized in the aluminum and steel industries and as a source of fluorine. It has been deposited from hydrothermal solutions in many places and is known from deposits of many geological ages.

Barite is a dense mineral that is semiopaque to X-rays and thus finds many uses in *x*-ray procedures used for medical purposes – for example, the white paste that people ingest before having their stomachs *x*-rayed is largely barite. It also finds use because of its density – for example, the density of barite is such that a slurry of barite mud is often used in oil drilling activities in order to prevent high-pressure gas from accidentally blowing the drilling liquid out of the well. Nearly all of the barite used for these and other purposes is mined from hydrothermal deposits.

ENERGY RESOURCES

The chief sources of energy now used on Earth are the fossil fuels (coal, oil, and gas), water power (including tides and waterfalls, both natural and at dam sites), biomass (especially wood), the sun, geothermal steam, the wind, and nuclear energy. Of these, only the source materials for nuclear energy are minerals, and only the fossil fuels of the coal family are rocks.

Nuclear energy minerals

In 1905, Albert Einstein demonstrated that matter and energy are equivalent and are related through his famous equation $e = mc^2$. Nuclear energy can be obtained from either very heavy or very light atoms. When a very heavy atom, such as uranium or thorium, is split into two lighter kinds of atom, energy is given off. *Nuclear fission,* as the splitting process is called, is the process that occurs in atom bombs and in all nuclear power plants built to date. When two atoms of a light element, such as hydrogen, helium, or lithium, are fused together to make a heavier element, a tiny bit of the mass of each atom is converted to energy. Such release of energy by *nuclear fusion* keeps the sun hot. Nuclear fusion may become a major source of energy in the future. If production procedures can be perfected, nuclear fusion very likely will meet the world's energy requirements for all time. It has been calculated, for example, that the fusion energy that could be produced from the top 0.001 in. of the oceans (i.e., a volume of seawater about equal to three months' discharge by the Mississippi River) would meet North America's anticipated energy needs for 10–20 thousand years. Furthermore, only relatively small amounts of radioactive waste would be produced! Unfortunately, no practicable method for producing nuclear fusion has yet been discovered. Until one is, minerals required for nuclear fission will continue to be sought.

Uranium is the most widely used of the natural fissionable materials. Deposits now supplying uranium minerals in North America comprise secondary concentrations in sedimentary rocks (Colorado Plateau region), hydrothermal vein deposits (Great Bear Lake region, Northwest Territories), and disseminated grains of unproved origin in metamorphosed sedimentary rocks (Blind River District, Ontario).

Coal

As noted in Chapter 4, coal consists largely, if not wholly, of macerals. In the classification we use in this book, all grades of coal are considered nonmetallic deposits of biological derivation. Each of the main members of the coal family – lignite, bituminous coal, and anthracite – represents a different degree of diagenesis that resulted in a loss of volatiles and a corresponding concentration of carbon. Coal beds occur as layers in stratified rock sequences, the precursor sediments of which were deposited in swamps (Fig. 6.11).

THE FUTURE

There is an old saying among miners and geologists that we can never know what the total resources of a given ore mineral are until the last grain has been mined. In essence, we shall not know which grain is the last grain until all of the Earth's crust has been dug up. That, of course, will never happen. That statement does not mean, however, that we shall always be able to find new ore deposits when old ones are worked out. So what can be predicted?

The search for new ore deposits has become an increasingly demanding and ever more technological activity. The day of the prospector and mule is long past. Today's prospectors fly over the ground measuring sensitive differences in the magnetism of rocks, put electrodes in the ground to detect subtle differences in electrical properties, and employ all sorts of sophisticated instruments, techniques, and means of analysis. For example, techniques have even been developed for rapidly measuring trace amounts of many elements and for detecting the tiniest levels of radioactive elements. In addition, when a modern prospector suspects that an ore body may be a few hundred or even a few thousand feet below the surface, special drills with diamond bits are used to bore small holes and

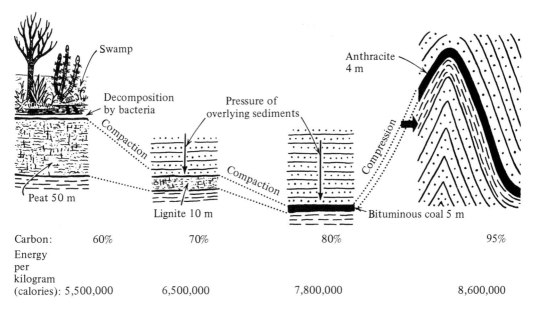

Figure 6.11. Conversion of plant matter from peat accumulated in a swamp to coal. Here, the changes in carbon content and the available energy per unit volume of material are correlated with the diverse products formed as the result of different amounts of compression and compaction of a 50-m-thick layer of peat, eventually to form a 4-m-thick layer of anthracite.

extract samples for direct examination. Indeed, prospecting has become much more a science and less an art, even though, as any geologist involved in mineral exploration will attest, experience and a "nose for ore" are still important.

Finding a new ore body today can cost tens of millions of dollars, and the testing of a newly discovered ore body, prior to actual mining, can cost hundreds of millions of dollars more. Nonetheless, the search for new ore bodies is continuous and for the most part has been extremely successful during the past several decades.

Important questions, however, remain – questions that can be answered only in the future. For example, a century ago Europe was the scene of a very active mining industry. Indeed, it was a net exporter of a number of mineral products, such as tin, copper, and mercury. Today, Europe is a net importer. The reason is that Europe's ore deposits have been worked out faster than new ones have been discovered. Meanwhile, the United States, Canada, South Africa, Australia, Chile, Brazil, the USSR, and a few other countries have become mining giants. Already, though, it is possible to detect shifts in the outputs of some of these countries. For example, fifty years ago the United States was self-sufficient in most mineral resources; now, year by year, it has to import more and more of the materials needed to satisfy its manufacturing needs. That is to say, today the United States is becoming more like Europe, an importer, and less like Australia, an exporter.

Is there a rhythm in the mining industry? Will it be necessary, in order for modern society to have the mineral resources it needs to continue its high living standards, for new lands to be explored more extensively and their mineral resources found and used? In contemplating the answers to these questions, it must be remembered that the size of the Earth is finite. Thus, as time passes and the population grows, a point will be reached at which there will be no new lands left on Earth to prospect. And it is hardly a cause for jubilation that that day is probably several decades in the future. Exactly how our human society will cope with the future resource crisis remains to be seen.

Some technologists believe that we will learn to mine leaner and leaner ores, all the way down to common rock. Possibly so, but many geologists are skeptical about such a conclusion. To them, it seems unreasonable to conclude that just because we can treat a copper ore and selectively recover copper, we also can mine a common rock consisting of silicate minerals and trace amounts of copper to recover only copper. Among other things, in the treatment processes that would be required, much larger quantities of iron, aluminum, magnesium, and a few other relatively abundant metals would also be produced, and surely would not be thrown away. Thus, many who have thought about the resource problems of the future have concluded that the result will much more likely be a society that will learn to use the most abundant metals – iron, aluminum, magnesium, manganese, silicon, and titanium – and also learn to do without the less abundant metals, such as copper, lead, zinc, gold, silver, and tin. Of course, there is always the chance that humankind's predilection for destroying civilization will make these predictions presumptive. On the other hand, perhaps medical research will permit many of us to wait and see what the future will bring.

FURTHER READING

Bates, R. L., 1960, *Geology of the Industrial Rocks and Minerals*. Harper, New York, 459p.
An introductory but nevertheless authoritative text. Despite its publication date, it is not much out-of-date.

Brobst, D. A., and Pratt, W. P. (editors), 1973, *United States Mineral Resources*. U.S. Geological Survey, Professional Paper 820, 722p.
An inventory and discussion of resources and reserves of all forms of nonenergy mineral resources.

Craig, J. R., Vaughan, D. J., and Skinner, B. J., 1988, *Resources of the Earth*. Prentice-Hall, Englewood Cliffs, New Jersey, 395p.
A comprehensive discussion of the geology, history, and environmental aspects of the production and use of all classes of mineral resources.

Evans, A. M., 1987, *An Introduction to Ore Geology* (2nd edition). Blackwell Scientific, Oxford, England, 358p.
A well-written paperback volume that is an introduction to all kinds of mineral deposits.

Guilbert, J. M., and Park, C. F., Jr., 1985, *Ore Deposits*. W. H. Freeman, San Francisco, 985p.
A detailed and authoritative treatment of all aspects of the geology of metallic mineral deposits.

Haymon, R. M., and Macdonald, K. C., 1985, The geology of deep-sea hot springs. *American Scientist*, 73, 441–9.
A summary of information about "smoker vent" springs.

Skinner, B. J., 1986, *Earth Resources* (3rd edition). Prentice-Hall, Englewood Cliffs, New Jersey, 184p.
An inexpensive paperback written at a high school senior level, covering all forms of mineral resources.

Teller, E., 1979, *Energy from Heaven and Earth*. W. H. Freeman, San Francisco, 322p.
Treats the origins and sources, uses, prospects, and plans and policies that relate to energy.

Tylecote, R. V., 1976, *A History of Metallurgy*. The Metals Society, London, 182p.
The only comprehensive book in English on this subject. Authoritative and very reliable. Well written, with extensive illustrations.

United States Bureau of Mines, 1985, *Mineral Facts and Problems, 1985 Edition*. U.S. Bureau of Mines, Bulletin 675, 956p.
This latest edition of this quinquennial publication summarizes information about, for example, the production, uses, economic factors, and supply and demand outlook for most mineral resources.

World Energy Conference (10th, Istanbul, Turkey, 1977), 1978, *World Energy Resources, 1985–2020*. IPC Science and Technology Press, Guilford, England, 249p.
An assessment of the magnitudes of all forms of energy resources by a group of international experts.

PERIODICAL

Economic Geology. Professional journal that is published eight times per year as the bulletin of the Society of Economic Geologists, P.O. Box 637, University of Texas–El Paso, El Paso, Texas 79968–0637.

7

Building materials

Rock materials were used for construction long before the first human beings appeared on Earth. Colonies of algae began to incorporate tiny grains of either calcite or aragonite (two polymorphs of $CaCO_3$) into their homes, which are called stromatolites, more than 3 billion years ago (Plate 40). Organisms ranging from one-celled foraminiferids to multicelled jawfishes have used mineral and rock fragments to build or enhance their dwelling places for nearly 500 million years. Insects such as ants and the larvae of caddis flies (Plate 41) have been building structures largely of sand and small pebbles for protection and living quarters for well over 200 million years.

Our human ancestors' use of rocks and rock materials for construction is a very recent development when compared with the just mentioned activities of some of the Earth's lower life forms. Nonetheless, the use of stones also goes far back into the dimmest recesses of human history. We know, for example, that our ancient ancestors used stones as tools and weapons and utilized large boulders and caves for simple shelters more than 2 million years ago. And we know that the earliest known paintings on the rock walls of caves date back about 30,000 years, that structures contrived to affect humans' aesthetic senses, as well as to provide shelter, have been built for the last 8,000–9,000 years, and that the Great Pyramid of Cheops, built some 5,000 years ago, was already a monument of antiquity when Moses lived "way down in Egypt land" (Fig. 7.1).

Today, mineral- and rock-based building materials contribute more dollars to the annual economy than the combined total for all other mineral resources, save the "mineral fuels." We describe them in two groups:

1. *building stones,* which are used just as they are taken from the earth (i.e., they are utilized after no treatment other than shaping or sorting), and

Building materials

(a)

(b)

Figure 7.1. Ancient stonework. ***(a)*** *The Sphinx and Pyramids at Giza, Egypt, which are some 5,000 years old, are made in part of large blocks of a fossiliferous limestone that was readily available within the region where they were built. The construction of pyramids such as this one is said to have involved some 100,000 to 200,000 workers for up to twenty years. Remarkably, the outer casing stones were apparently laid from top to bottom – that is, the workers began laying them at the peak and ended with the lowest course. (Photograph by R. V. Dietrich)* ***(b)*** *These paving stones, laid more than 2,000 years ago on the Appian Way, south of Rome, are the fine-grained igneous rock basalt. This rock type, as crushed stone, is still considered to be and is used as one of the most desirable rocks for highway construction. (Photograph by B. J. Skinner)*

2. *rock products,* which are produced by firing or melting of crushed minerals or rocks.

BUILDING STONES

Dimension stone

The term widely applied to rock sold in blocks or slabs of specified sizes and/or shapes, and also for rough boulders that are used in construction (Plate 42), is *dimension stone*. Dimension stones can be used rough, or they can be trimmed, chipped, rough cut, cut and rubbed, or cut and polished (Plates 43 and 44).

Diverse dimension stones are used for all sorts of construction, monuments, and sculptures and also for paving and curbing. In buildings, they are used for foundations, walls, and roofs and for both exterior and interior facing and trimstone. Several rocks used as dimension stone are listed in Table 7.1.

Roughly dressed slabs of granite were used for both paving and wall

Table 7.1. *Rocks used as dimension stones*

Rock category[a]	Rocks (examples only)
Igneous	"Granite,"[b] syenite (larvikite), diorite, gabbro, and dolerite (= diabase = "traprock"); rhyolite, trachyte, andesite, and basalt; obsidian; diverse porphyries
Pyroclastic	Ash tuff (e.g., peperino) and ignimbrite
Sedimentary	Limestone (e.g., "oolitic" and numulitic), travertine, and chalk; sandstone (e.g., "brownstone" and "freestone"), puddingstone (e.g., "jasper"), and conglomerate (e.g., "calico rock"); gypsum (alabaster)
Diagenetic	Dolostone and recrystallized limestone (e.g., "Tennessee marble")
Metamorphic	Marble;[c] gneiss, amphibolite, schist, and slate; serpentinite, soapstone, and greenstone
Migmatitic	Migmatite (e.g., "Morton gneiss")
Other	Laterite, duricrust

[a] Categories are described in Chapter 4.
[b] *Granite* as a term is frequently used in the marketplace to include igneous rocks that geologists identify as granodiorite, diorite, and syenite, as well as granite per se. Such terms as "black granite" may include additional igneous rocks such as gabbro and diabase.
[c] *Marble* as a term is frequently used in the marketplace to include sedimentary and diagenetic limestones and dolostones that take good polishes, as well as marbles per se (in petrology, "marble" is the accepted name for *metamorphic* rocks consisting largely of calcite and/or dolomite). The combination of all "marbles" of the marketplace is estimated to account for more than 50 percent of all building stone used to date.
Source: Based in part on Kempe (1983).

linings in Egypt during the first dynasty (~3100–2900 B.C.). The well-known architecture historian Norman Davey has written that "possibly the earliest complete building of stone that can be dated is the pit chamber of the tomb of King Khasekhema... at Abydos [Egypt], of 2nd dynasty date [i.e., ~2900–2700 B.C.]." Other than prevailing taste, which usually is dictated by architects, the characteristics that usually determine which stone is used for any given purpose are availability, appearance, durability, and workability.

The availability of dimension stones has frequently been cited as the primary control of architectural styles that characterize certain regions. As the famous English writer John Ruskin noted in his *Stones of Venice* (1887), "all over the Roman empire [they] set to work, *with such materials as were nearest at hand,* to express and adorn herself as best she could" (italics added). Availability is also emphasized by the fact that some building stones have been used more than once. For example, travertine blocks used during the 1st-Century construction of the Colosseum (Amphitheatrum Flavium) in Rome were later salvaged from its ruins, during the 15th and 16th Centuries, and reused in St. Peter's Church in Vatican

City and for several other buildings in Rome. (No wonder the Colosseum we view today is merely a remnant of its once much larger and more impressive self.)

Nonetheless, some dimension stones have been transported over relatively long distances:

Four thousand years ago, the giant monoliths of Stonehenge in England were dragged and rafted many tens of miles.

More than 100,000 Devonshire slates were shipped from southwestern England to Mont-Saint-Michel, in France, in 1436.

Since the late 1800s, several stones, such as larvikite from Norway and marbles from Italy, have been transported the world over for use as facing and trimstones.

Appearance depends not only on a stone's color and texture but also on such things as the kind of surface the stone is to be given. Fortunately, the diversity of rocks affords an almost infinite number of possible choices. Available colors range from black to white and include shades of all of the colors of the spectrum. Some stones are best polished, others are best merely sawn, and still others are best left rough.

Durability depends on such characteristics as the minerals present and the porosity and permeability of the overall rock. In order for a rock to be considered for outdoor use, it must be able to resist deterioration, including staining, when exposed to the weathering agents that are active within the area where the rock is to be placed. Today, it is recognized that one must even consider such things as the potential effects of acid rain.

Workability includes such aspects as whether or not a rock exhibits a good contrast between polished and tooled surfaces and whether or not it can be easily carved (Fig. 7.2). It is because of workability requirements that dimension stones must have more or less uniform textures, as well as an absence of closely spaced fractures or planes of weakness, if they are to be used for certain purposes. For example, textural differences might distract from statues, especially if the differences were suggestive of such things as blemishes on surfaces representing human skin.

A rock used as a building stone must also be strong enough to support the weight of the overlying structure. Strength is a property that can be checked by laboratory tests; in practice, however, this property is seldom tested, because it is well known that just about any rock that might be considered for use where strength is important would far surpass any requirements that might be set. In fact, it has been calculated that a building made up of any "average" dimension stone would have to be nearly 2.5 mi high before there might be failure of the lowest courses attributable to overload.

Gems, granites, and gravels

Figure 7.2. The statue of Abraham Lincoln. This famous statue, sculpted by Daniel Chester French, consists of Georgia marble. It is on a base of Tennessee marble (actually a diagenetic, recrystallized limestone). The main structure of the memorial in which the statue is located is made of Yule Marble from Colorado, has columns and lintels of Indiana limestone, foundation steps of Massachusetts granite, and floors of Tennessee marble, all blended into what has been termed an architectural classic. (Photograph by R. V. Dietrich)

Although many dimension stones are extremely durable and nearly all provide superior insulation, most present-day uses are based on custom or aesthetic desire. Less expensive substitutes, such as blocks and slabs of concrete or bricks and glass, are usually used in lieu of more costly natural stones. Therefore, as might be expected, dimension stone production has been decreasing for at least the last four or five decades. In fact, several producers of dimension stone have found it impossible to operate at a profit without also producing one or more salable by-products, such as crushed stone (Fig. 7.3).

Crushed stone

Rock that has been broken into smaller, typically irregularly shaped fragments is called *crushed stone*. The fragments, usually sorted by size,

Figure 7.3. Granite quarry, Mt. Airy, North Carolina. This is one of the largest open-faced quarries in the world. Original production was almost exclusively dimension stone – that is, quarried stone cut, and in some cases polished, for use as finished stone paneling or curbing. In order to continue operation at a profit, current production also includes such things as crushed stone, chips for landscaping, and turkey grit. (Photograph by R. V. Dietrich)

may range from fine powder up to blocks a few feet in greatest dimension; the latter usually are called *riprap* (Plate 45). Most crushed stone is made from blocks that are quarried, crushed, and screened or otherwise separated into the desired size fractions.

So far as we know, crushed stone has been used only since the mid-1700s, when the French civil engineer Pierre Tresaguet developed a system of highway construction that used a topping of walnut-size broken rock fragments. Early 19th-Century macadam laid in England constituted another fairly early use of broken stone.

Until the late 1850s, crushing was done by hand sledging – often by convicts (the Sing Sing method). At best, this production was inefficient, primarily because it tended to yield fragments in a wide range of sizes rather than of some uniform size range as is usually desired. In 1858, Eli Whitney Blake pondered these problems and invented the first mechanical rock crusher in time to help fill orders for the large quantities of crushed stone required for a two-mile-long macadam road that was being laid between New Haven and Westville, Connecticut. Blake's jaw crusher was the prototype of crushers now used in quarries around the world.

During the century following Blake's invention, the crushed stone industry of the United States grew to include more than 4,000 quarries, with an annual production of more than 500 million short tons (a short ton = 2,000 lb) and a value of nearly $750 million. More recently,

A MULTIROCK MONUMENT

The Fountain of Freedom monument, soon to be built on Independence Mall in Philadelphia, Pennsylvania, will include blocks of rock from each of the United States (Fig. 7.4). Rocks from all of the families are represented. There are several granites, anorthosite, diabase, and basalt; several limestones (including a reef rock), sandstone, dolostone, and petrified wood; diverse marbles, quartzite, gneiss, and a migmatite; and both a gold ore and a silver ore. The rocks range from less than 10,000 to more than 3.5 billion years old. Indeed, considering the way the rocks were selected, one might suggest that Providence intervened so that rocks of many kinds and many ages would be included, as if to reflect the polyethnic makeup of the United States's citizenry.

The project that led to the collection of all these rocks and the building of this monument was conceived by **We the People 200,** a nonprofit organization that was created "to provide a lasting legacy to the Bicentennial celebration of the U.S. Constitution." State governors were contacted, and the "wheels were set in motion." Subsequently, several federal and state agencies and their personnel and, in some states, other individuals, organizations, and companies became involved in the efforts that resulted in the selection, preparation, and transportation of the rocks to Philadelphia. The delivered blocks ranged up to 3 × 4 × 8 ft in size and up to approximately ten tons in weight.

The transportation to Philadelphia of the rocks contributed by California, Utah, Wyoming, Colorado, Montana, North and South Dakota, Missouri, and West Virginia became the theme of a documentary (**Rock Across America**). They were carried by a caravan that took a reverse journey along the Oregon Trail from the California gold fields to Independence, Missouri, the frontier trailhead for many forty-niners, before continuing more or less directly to Washington, D.C., and then on to Philadelphia. Elaborate dedication ceremonies were held as each rock was added to the caravan. State governors made speeches, and parades were organized; there was bell ringing and waving of flags, and in one place cancan dancers emulated their 19th-Century sisters.

Figure 7.4. Model of pylons for the Fountain of Freedom, soon to be built on Independence Mall in Philadelphia, Pennsylvania. The arrow points to the surface of one of the pylons to be faced by the rocks sent by the states. (Courtesy U.S. Department of the Interior, National Park Service)

Each of the rocks has a story – both geological and otherwise. A few examples follow:

The California contribution is a gold-bearing metamorphic rock from Sutter's Mill, where the gold nugget was found that started the 1849 gold rush.

Colorado sent a piece of one of the most famous mountains in the world, the aforementioned 3 × 4 × 8-ft block of Pike's Peak Granite. It was while on the summit of this mountain that Katharine Lee Bates was inspired to write "America the Beautiful."

The Delaware block, "Brandywine Granite," was quarried in 1802 along with blocks that were used for the foundation of Du Pont's original black powder mill.

Florida's rock, the Key Largo Limestone, is a reef rock that is composed largely of the skeletal remains of corals, mollusks, and algae, along with lime sand.

The Louisiana limestone is a piece of the caprock from one of the salt domes. It had to be obtained from a mine, because Louisiana has no natural rock exposures.

Michigan's Kona Dolomite is renowned because of the attractive appearance it assumes when cut and polished. It is often used for jewelry and diverse decorative pieces.

Montana's travertine is thought to represent post–Ice Age deposition of calcium carbonate by a hot spring.

The North Dakota contribution, a large granite boulder taken from the grounds of the state capitol, was almost certainly transported into North Dakota from the Canadian Shield during the Pleistocene Ice Age. Therefore, the monument includes a piece of rock representing the bedrock of Canada as well as rocks that constitute the bedrock or occur in subsurface sequences beneath the United States.

Oregon's rock, a volcanic lava, contains crystals of the official state gemstone, a variety of feldspar sometimes called "sunstone."

"Tennessee marble" probably has been seen by as many people as any rock ever marketed. (Polished slabs of the rock have long been used as dividers in public rest rooms.) This rock is the coarse-grained, commonly fossiliferous rock that exhibits dark gray to brown graphlike lines called **stylolitic seams.**

Several of the granites, marbles, and sandstones and Minnesota's migmatite have had long histories of use in the building trade, especially in North America. They have been used for such things as monuments, exterior and interior ashlar and trimstone, and interior wainscoting. To mention just a few examples:

Vermont's Barre Granite, from the famous Rock of Ages Quarry, is one of the most used monument stones in the world.

Indiana's Salem Limestone (widely marketed as "Indiana Limestone") and Minnesota's Morton Gneiss (a migmatite with red, black, gray, and nearly white hues arranged in rather flamboyant patterns) may be seen as the trimstones and ashlar on thousands of buildings throughout North America.

A few of the many well-known edifices in which rocks represented in this memorial have been used include the new AT&T Building in New York City (Connecticut's Stony Creek Granite), the foundation of the Jefferson Memorial in Washington, D.C. (Georgia's Elberton Granite), the Southern Pacific Building in San Francisco (Idaho's Boise Sandstone), the reflecting pool in front of the Lincoln Memorial in Washington, D.C. (North Carolina's Mount Airy "Granite"), and both Grand Central Station in New York City and the National Art Gallery in Washington, D.C. ("Tennessee marble").

As one might expect, some rock and mineral collectors have begun to collect pieces of the same rocks that are included in this monument. The resulting specialized collections will, we feel sure, be exhibited with pride at club meetings and shows. This pursuit seems quite worthwhile, especially if the exhibitor personally collects the rocks and takes the time to study each rock in order to learn its geological history so that it can be communicated to future viewers of the collection.

---

annual production figures for the mid-1980s reached nearly a billion short tons, with a value of more than $3.75 billion.

A large percentage of crushed stone is used in the construction industry. It is utilized directly as aggregate for both concrete and bituminous mixes, as road metal and railroad ballast, as riprap for breakwaters and both shore and highway-cut protection, and as stucco dash and terrazzo chips; it is used indirectly as the raw material for the production of components made of, for example, rock wool. In addition, crushed stone has an incredible diversity of other uses – for example, as agricultural lime (fertilizer), blast-furnace flux, turkey grit (to aid the fowl's digestion within its crop), sewage filtration beds, inert ingredients in drugs, and raw material for the manufacture of several chemicals.

Limestone and dolostone account for well over half the annual production of crushed stone. This predominance depends on two facts: (1) both of these rocks meet the strength specifications set up for most uses, and yet (2) both are made up of relatively soft minerals that are not so hard on crushers and other equipment as most other rocks are. Consequently, they are less expensive to produce than most other rocks. Granitic rocks, diabase and basalt (often termed "traprock"), and sandstone are the other rocks used rather widely. Primarily because of their ready availability, still other rocks, including such diverse kinds as volcanic cinders (scoria) and vein quartz, are used in a few places, usually where the more commonly utilized rocks do not occur nearby.

Stringent specifications have been established for many uses of crushed stone. The American Society for Testing and Materials (ASTM) is continually reviewing and publicizing the properties that need to be taken into account for diverse uses. The specifications relate to such things as the shapes of the fragments, sizes and size distribution of fragments, and both their physical and chemical properties. The pertinent physical properties control requirements relating to strength, toughness, resistance to abrasion, porosity, and absorption. The chemical properties usually considered deal with the stability of mineral constituents whose presence might lead to relatively rapid weathering or to reactions with certain kinds of cements.

Although most rock used for crushed stone is quarried from open pits, some of it is mined underground. In some places the underground mining is dictated by environmental considerations or zoning restrictions; in others it has been found to be economically advantageous – for example, in areas where weather conditions preclude year-round operation of open pits.

The production of crushed stone probably will continue to increase for many years into the future, especially for use as aggregate for concrete and bituminous mixes.

Building materials

(a) (b) (c)

Figure 7.5. Concrete often spalls and/or becomes stained when reactions take place between the cement and the atmosphere and/or the aggregate. Such problems can often be reduced by establishing sufficiently stringent specifications for the aggregate. (a) A concrete sidewalk in which the sand and gravel contained clay masses and iron-rich concretions. (b) Spalling caused by absorption of water by a clay ball and consequent expansion. (c) Rust-stained streak emanating from an iron-rich concretion. (Photographs by R. V. Dietrich)

Sand and gravel

As described in Chapter 5, sand and gravel are unconsolidated aggregates of sand grains and of sand grains plus pebbles, respectively. As already mentioned, most sand consists largely of grains of the mineral quartz. The pebbles of gravels may be of only one, a couple, a few, or several kinds of rocks.

Specifications established for sands and gravels differ with use and sometimes with such things as an architect's preferences or even the character of the sands or gravels that are readily available. For construction purposes, the "specs" are similar to those mentioned for crushed stone – that is, they relate to strength, resistance to abrasion, and chemical stability (Fig. 7.5) and to particle shape, size, and size distribution. For a few uses, such as children's sandboxes, cleanliness – that is, absence of dust – is also desirable.

In the United States, the bulk of sand and gravel is used as roadbase for highway and street construction and for fill. More than a third of the sand and gravel produced, however, is used for aggregate. Smaller quantities are used as foundry and molding sands, railroad ballast, abrasives, and septic field filtration media and for snow and ice control. Even smaller quantities are consumed in the manufacture of glass, certain chemicals, and alloying agents.

The amount of sand and gravel available is said by some economic geologists to be essentially inexhaustible, and strictly speaking this is true. This statement is, however, misleading, because within many of the areas where sand and gravel are needed most – for example, near the world's big cities – deposits are no longer readily available for local production. This is so because local deposits, if they ever were present, have been greatly depleted, have been built on, or are no longer accessible because of environmental considerations and/or zoning restrictions precluding quarry operation.

ROCK PRODUCTS

Cement and concrete

As the following interdependent definitions indicate, cement and concrete are intimately related.

Cement is the substance that, along with admixed aggregate and water in appropriate proportions, will harden with time (i.e., "set") to form concrete.

Concrete is the name applied to rocklike materials that are produced when appropriate proportions of cement, rock and/or sand and gravel aggregate, and water are mixed and allowed to react and set. The setting is, in essence, crystallization to a mass of interlocking grains.

Mortar made from cement differs from concrete in that its aggregate consists only of sand. In construction, concrete is poured and allowed to cure in place or is used as precast forms that are fabricated at cement plants, transported, and then bound together with mortar at the construction sites.

Lime mortars were used in ancient Egypt at least 5,000 years ago. Those mortars consisted of quicklime (CaO) that was mixed with water and sand and then allowed to harden. The quicklime usually was produced by heating limestone ($CaCO_3$) to approximately 900°C (~1,650°F) to drive off the carbon dioxide (CO_2).

A siliceous (i.e., silica-rich) cement called pozzuolana cement was used by the Romans as early as 100 B.C. It was made by stirring a mixture of quicklime and powdered volcanic glass (obsidian) from Pozzuoli, which is near Mount Vesuvius, together with water and then letting the mixture solidify. The quicklime and water reacted with the volcanic glass to form interlocking needles of calcium silicate. With the decline of the Roman Empire, however, the use of such cement disappeared until the middle of the 18th Century.

During the last half of the 1700s, two Englishmen, John Smeaton and James Parker, rediscovered cement making. Smeaton, the engineer who designed the famous third Eddystone Lighthouse, which is on the English Channel about 14 miles off Plymouth, is said to have found the Romans' cement-making formula while examining an ancient Latin document. Parker patented a process that involved the crushing and heating of natural "cement rock" to produce a glassy mass and then grinding of the resulting clinker into a powder, which he called Roman cement. (Natural cement rock is an impure rock that has the overall composition of the different rocks and/or other substances that usually are crushed and mixed to manufacture cement.)

Some two and a half decades later, in 1824, another Englishman, Joseph Aspdin, produced the first Portland cement. He made it by powdering a fired mixture of limestone and clay. Aspdin's cement was named for the Isle of Portland, which is south of Weymouth on the north side of the English Channel, because the cement, when hardened, resembled the widely used dimension stone that was quarried there. (Incidentally, the Portland stone has been used for such well-known buildings as St. Paul's Cathedral in London and the United Nations Building in New York City.)

In the United States, cement was first made in 1818; that cement, a Roman cement, was produced by an engineer named Canvass White from a natural cement rock that occurs near Chittenango in central New York. Portland cement was first produced in America in 1872, when David Saylor opened a small plant at Coplay, in eastern Pennsylvania. It was thirty-seven years later when the first mile of rural concrete pavement was laid in the United States; it was put down in what is now Detroit, Michigan.

Portland cement accounts for a very large percentage of all the cement produced today. The raw materials usually used are limestone and shale. Other rocks or substitutes – such as oyster shells, sand, clay, and slag – are, however, also used rather widely. The correct proportions for the furnace feed are 60–65 percent lime (CaO), which comes from the limestone or oyster shells, plus 10–25 percent silica (SiO_2) and 5–10 percent alumina (Al_2O_3), both of which come from the shale and/or a combination of other ingredients. The raw materials are crushed to powder, mixed, and fired at approximately 1,480°C (~2,500°F); the clinker thus

Gems, granites, and gravels

Figure 7.6. The Sydney (Australia) Opera House is a fine example of modern-day use of concrete. The famous Sydney Harbour Bridge is in the background. (Photograph by R. V. Dietrich)

produced is mixed with 4–5 percent gypsum, which serves to control the rate of hardening; the mixture is then pulverized to a powder, which is the Portland cement. For some specific uses, either the original mixture or the final mixture or both can be modified so that the cement will, for example, resist acids. Obviously, concrete made from such a cement will last much longer than common cement when used for foundations emplaced where acid soil waters prevail.

Today, concrete comprises about 70 percent of the total of all the structural rock materials used. This broad-scale use is obvious when one looks at the makeup of our 20th-Century buildings (Fig. 7.6), dams, bridges, and highways.

Brick and tile

Brick and tile are both made from clay. To make them, moist clay, plus or minus sand and/or chaff and/or other additives that reduce shrinkage, is molded into the desired shapes and then hardened either by drying in the sun or firing in a kiln. If a glazed surface is desired, the molded forms are sprayed with a ceramic glaze and fired at about 1,090°C (~2,000°F).

Although most bricks currently being made are rectangular blocks, all sorts of custom shapes are also produced. Most tiles are thin slabs, curved pieces, or hollow tubes, and large percentages of them are glazed.

Vessels made of hardened clay date back to at least 25,000 B.C. Tablets inscribed with ancient cuneiform characters, monuments, buildings, walls, and pavements made of brick and/or tile were used at many places in the Middle East long before the Christian era. For the most part, those clay products were produced and used where natural rocks were scarce – for example, in areas underlain by floodplain deposits, such as those in the plain between the Tigris and Euphrates rivers, a region that today lies in Iraq.

The ancient city of Babylon had brick and tile palaces, temples, and other buildings and was encircled by a brick wall said to have been 26 m (~85 ft) high, nearly 100 m (~325 ft) thick, and about 15 km (~9.5 mi) in perimeter. Eighteen successive levels of sun-dried brick structures, dating back to the fourth millennium B.C., were found in the excavations at Eridu, the holy city of the Sumerians, in what is now southern Iraq. Many of the structures were made largely of sun-dried bricks, most of which contained rather large amounts of straw and/or hair, but were veneered by glazed, kiln-fired bricks or tiles. The veneers were apparently added because sun-dried bricks tend to disintegrate if wet, whereas fired bricks are irreversibly dehydrated and thus are not affected by wetting. Brick and tile structures were also made in ancient Egypt, sporadically throughout the Roman Empire, and in the ancient Orient (Plate 46). Some particularly interesting bricks were made in Rome during the 1st, 2nd, and 3rd Centuries A.D.; all fired bricks, they are of diverse colors, shapes, and sizes and were arranged so as to constitute some rather elaborate, multicolored cornices, capitals, and so forth.

Adobe, a kind of sun-dried brick, was used by American Indians even in pre-Columbian times – especially by Indians in what is now the southwestern United States. Bricks, other than adobe, were not used in America until 1650, when they were incorporated into some of the mansions on Manhattan Island, then New Amsterdam. Those bricks were imported from Europe – probably from Holland, but possibly from England.

Today, brick and tile production is a flourishing industry. Uses include the following: Bricks of diverse sizes and colors are used to face many buildings ranging from small houses (even doghouses) to multistoried condominiums, ornate churches, and large commercial complexes. Brick chimneys are standard in many areas. Though no longer used for paving, bricks as well as tile are still utilized widely for decorative walkways and patios. Fire bricks, which are highly refractory, are required as linings for fireplaces, ovens, kilns, and blast furnaces. Tile is used for drainage pipes, chimney linings, fireplaces, hearths, patio "blocks," and roofs.

Nearly all brick- and tile-producing plants are located near their sources of clay. The clay deposits are of several origins, including those formed by weathering in situ (i.e., residual deposits) and those deposited in marine, estuarine, lake, swamp, and stream environments. In some places, shale or slate, which are clay-rich rocks, are pulverized and used in lieu of unconsolidated clay.

Clay is not a single mineral; rather, it comprises a group of some twenty closely related minerals — for example, kaolin, dickite, montmorillonite, halloysite, and nacrite — all of which are hydrous aluminum silicates. Nearly all clay minerals are plastic when wet and become rocklike when subsequently dried or fired.

Both the texture and color of bricks and tile depend on the impurities in the parent clay or clay-rich rock and the temperature and chemical conditions — for example, oxidation versus reduction atmospheres — within the kilns in which they are fired. In some places, bricks and tiles of several colors — for example, red, orange, brown, and buff — are produced from the clay or clay-rich rock from a single pit or quarry. Bricks of these diverse colors are generally attributable to the iron content of the clay. However, manganese, titanium, and organic compounds may also be naturally present or added and may act independently or in concert with one another and/or with iron to modify the color.

Clays have, of course, several other uses. For example, they serve as the raw materials for terra-cotta, faience, pottery, and a few other ceramic wares, for lightweight aggregate, drilling mud, and absorbents for cleaning up oil, and for animal litter and inert fillers for such products as paint, paper, and rubber.

Plaster

Strictly speaking, the term *plaster* includes all the diverse substances used to coat and thus conceal rough or irregular surfaces. Materials used as plaster include mud, quicklime, cement grout, and plaster of Paris.

Some surfaces are plastered merely to make them suitable for decoration. Examples of ancient plastered surfaces can be seen here and there throughout the Middle East — for example, in several chambers in Egyptian pyramids. Two especially noteworthy examples of the use of plaster are the intricate precast panels made by the Moors for the Alhambra Palace in Granada, Spain (Plate 47), and the ceiling of the Sistine Chapel in the Vatican, on which Michelangelo painted his incomparable masterpieces.

Today, the term *plaster* almost always means plaster of Paris. Plaster of Paris is produced by heating crushed gypsum ($CaSO_4 \cdot 2H_2O$) to approximately 190°C (~375°F) in order to drive off part of the water and then pulverizing the product to a fine powder. Plaster per se is formed when that powder is mixed with an appropriate amount of water and allowed to harden. What really happens is that on heating, part of the water of the $CaSO_4 \cdot 2H_2O$ is lost, and $CaSO_4 \cdot \tfrac{1}{2}H_2O$ (hemihydrate) is formed; subsequently, the powdered hemihydrate – that is, the plaster of Paris – can be mixed with water, and the pasty mixture will dry out and crystallize to tiny interlocking grains of gypsum, which is the hardened plaster. The name plaster of Paris was given to this kind of plaster because large quantities of it were produced from beds of rock gypsum that underlie Paris, France.

Sometimes the pasty mixture of plaster of Paris and water is applied directly to the surface to be covered. More often, the plaster is sandwiched between two sheets of heavy paper, marketed as wallboard, and installed as "drywall." Particles of foreign materials are sometimes added as aggregate to the plaster of Paris to give the the hardened plaster some particular characteristic – for example, for use as high-sound-absorbency members for walls and ceilings of music rooms.

Present-day annual production is said to amount to more than enough plaster to coat 100 mi^2 of wall – that is, enough to coat an 8-ft wall that would extend from Washington, D.C., to Chicago, Illinois.

Glass

Manufactured glass, like natural glass, is a supercooled liquid and thus is amorphous. Manufactured glass, unlike most natural glass, is almost pure silica. For most purposes, the composition is closely controlled during production in order to obtain the desired characteristics – for example, the degree of transparency.

Cavemen used natural glass for tools and weapons. Pliny, in his *Historia naturalis,* written in the 1st Century A.D., recorded the story that manufactured glass was first made by accident by some Phoenician sailors who were cooking over open fires on a beach near Acra:

Once a ship belonging to some traders in natural soda put in here and . . . scattered along the shore to prepare a meal. Since, however, no stones suitable for supporting their cauldrons were forthcoming, they rested them on lumps of soda from their cargo. When these became heated and were completely mingled with the sand on the beach a strange translucent liquid flowed forth in streams; and this, it is said, was the origin of glass.

Whether true or not, records preserved in hieroglyphics and found in Egyptian tombs indicate that glassmaking was well established by 1500 B.C.

Glass appears to have been first used in construction when Romans used pieces left over from their mosaics to close openings in the walls of their dwellings. Although most of the glass they made for their mosaics was at best only translucent, that early use in walls appears to have led to glass windows by about 1000 A.D., when relatively clear glass was produced in Venice.

One especially well known use of glass in buildings was in the stained-glass windows in the Renaissance cathedrals of western Europe (Plate 48). Those magnificent rose windows were substitutions of glass for stone and, in addition, served to decorate and lighten the otherwise rather somber interiors of the structures. Another well-known use gained impetus late in the 17th Century, when Louis Lucas de Nehou produced large polished plates of glass that made especially fine mirrors, and Louis XIV had hundreds of the plates installed in the Hall of Mirrors in the palace at Versailles.

During the last few decades, glass has taken on a major role in construction. One needs only to look around in any modern urban area to see that many buildings erected during the last few decades appear from the outside to be made almost wholly of glass.

Glass, like brick and tile, is a ceramic material – that is, it is made by firing minerals and/or rock materials and allowing the product to harden. The general process of glassmaking involves the following steps: The chief raw material – silica-rich sand, sandstone, and/or quartzite – is ground to a powder, and if necessary purified; the powder is mixed with limestone and soda ash or other substances that provide lime (CaO) and soda (Na_2O) to lower the melting temperature of the silica; the mixture is melted; the melt is quenched – that is, it is cooled so rapidly that nucleation and crystallization are precluded. The quenched melt is glass.

In order to give glass special characteristics, certain elements or compounds can be added to modify the melt and thus the product. For example, lead (Pb) increases luster, lithium (Li) improves strength, potassium (K) increases thermal expansion, and several different elements and combinations of elements produce diverse tints and colors.

Production of glass mushroomed soon after World War II and has continued to increase ever since. Most of it is used in windows, including those in automobiles and other vehicles. Other uses include glassware and bottles, lightbulbs, television and other electronic display tubes, and the fiberglass and foam used for heat and sound insulation and for several other well-known purposes.

SOME FAMOUS STRUCTURES

It is generally agreed that the architecture characteristic of a culture has depended to a large extent on the kinds of construction materials that have been readily available near where they have been used. The massive

Egyptian pyramids, the exquisite Greek temples, the magnificent western European cathedrals, and even the steel–glass and aluminum–glass skyscrapers of this century (Plate 49) are examples. A few of our favorite structures that exhibit the utilization of diverse building stones and rock products are shown in Figures 7.1, 7.2, and 7.7 and Plates 43 and 45–48.

FURTHER READING

Bates, R. L., and Jackson, J. A., 1982, *Our Modern Stone Age*. Kaufmann, Los Altos, California, 136p.
Includes information on production and uses of a few rocks.

Davey, N., 1961, *A History of Building Materials*. Phoenix House, London, 260p.
A well-illustrated, well-documented, extensive – though somewhat dated – history of nearly all of the common building materials.

Eichholz, D. E., 1962, *Pliny Natural History* (volume X). Harvard University Press, Cambridge, Massachusetts, 344p.
The quotation about the discovery of how to manufacture glass is on page 151. Several additional interesting items about minerals and rocks, as viewed during the early days of the Roman Empire, are also covered in this volume.

Kempe, D. R. C., 1983, The petrology of building and sculptural stones, pp. 80–153 in Kempe, D. R. C., and Harvey, A. P. (editors), *The Petrology of Archaeological Artefacts*. Clarendon Press, Oxford, England.
A well-documented summary of the diverse rocks used in ancient structures. It does, however, contain several errors in petrographic nomenclature.

Lefond, S. J. (editor), 1983, *Industrial Minerals and Rocks* (5th edition). American Institute of Mining, Metallurgical, and Petroleum Engineers, New York, 1477p.
The most recent edition of a widely used reference concerning industrial minerals and rocks.

Pliny: See the Eichholz entry.

Ruskin, J., 1887, *The Stones of Venice*. Wiley, New York, vol. 1, 434p., vol. 2, 397p., vol. 3, 376p.
A rambling, overly wordy, overly long three-volume work that nonetheless contains all sorts of information and interpretations about southern Gothic and other architectural forms in Venice and vicinity – the quotation in our description is on page 15 of volume 1.

United States Department of the Interior Geological Survey, 1975, *Building Stones of Our Nation's Capital*. U.S. Government Printing Office, Washington, D.C., 44p.
An especially well illustrated color booklet about rocks used in buildings, memorials, houses, fountains, and a bridge in the District of Columbia. Thirty-nine buildings and other stone structures of interest are shown on a map and described individually.

8

Rocks and minerals in diverse environments

PLATE TECTONICS

Starting in the 1950s, and continuing to the present, discoveries have led to a startling new understanding of the fundamental way our Earth works and why. Now we know why the Alps (Plate 50), the High Sierra, the Cascade volcanoes, the Andes, and the Hawaiian Islands are where they are. This new "knowledge," which might better be termed a revelation, is referred to as *plate tectonics*.

The word "tectonic," as used by geologists, pertains to the study of the structure of the Earth's crust and the forces that produce changes in its structure. The word is derived from the old Greek word *tekton*, meaning a carpenter or builder. The plate tectonic paradigm stems from the discovery that an outer layer of our Earth, down to a depth of about 100 km (~60 mi), consists of huge fragments – called plates by geologists – that move and produce the geological structures that are manifested by, for example, the Earth's mountains and ocean basins.

The brittle outer layer of Earth is called the *lithosphere* (from *litho*, meaning rock (Fig. 8.1). It is the lithosphere upon which we live – the part of Earth that we can see and sample. It floats on an easily deformed, fluidlike plastic layer called the *asthenosphere* (from the Greek combination form *a-*, meaning without, and *sthenos*, meaning strength). The interplay between these two layers determines where ocean basins and continents are located, where volcanoes occur, where mountain ranges form, and, most important so far as the topic of this book, where diverse minerals and rocks are formed.

When solid materials such as iron, copper, glass, or even rock are heated sufficiently, their strength drops. This kind of decrease explains the presence of the asthenosphere below the lithosphere: The Earth's interior is extremely hot; temperature rises steadily with depth, from about 20°C (~70°F) at the surface to thousands of degrees at the core.

Rocks and minerals in diverse environments

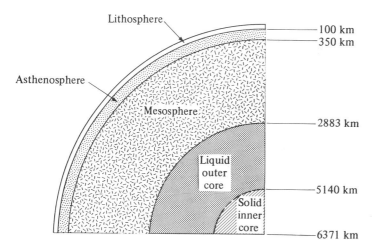

Figure 8.1. Layering of physical properties in the earth. Changes are due to changing temperature and pressure with depth. The lithosphere is hard and rigid as compared to the asthenosphere, which is a region where rock is weak, plastic, and easily deformed. The mesosphere is stronger than the asthenosphere because even though temperature increases with depth, increased pressure offsets the effects of temperature.

Because of this temperature gradient, usually called the *geothermal gradient,* the strength of the rock materials drops with depth, and it does so dramatically at the lithosphere–asthenosphere boundary. In fact, the changes in properties at that boundary can be easily sensed and charted by seismic – that is, earthquake – waves.

The Earth's internal heat appears also to be responsible for the movement of the lithosphere plates. Heat escapes steadily to the surface of the Earth and is radiated out into space. This happens because heat always flows from hot toward cooler loci. Heat moves in two ways – one slow and inefficient, the other more rapid and fairly efficient. The first, *conduction,* is the way heat moves up the handle of a hot saucepan; the second and far more efficient way is by *convection.* Thus, as might be suspected, convection is the primary method of heat transfer in Earth materials.

When any mass of material – solid, liquid, or gas – is heated, it expands, and consequently its density decreases. As a result, if a hot, less dense object is in a cool, more dense environment, the less dense object will tend to rise and float upward. Within the asthenosphere, and even deeper within the Earth, some regions are apparently hotter than others. Thus, great convective motions stir deep within the Earth (Fig. 8.2), and those movements appear to be responsible for the motions of the overlying lithosphere. That is, convective motions within the Earth appear to

Gems, granites, and gravels

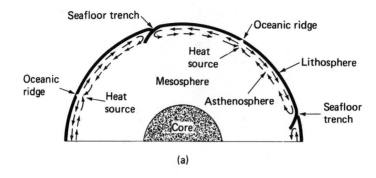

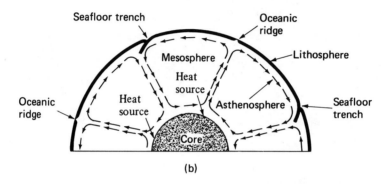

Figure 8.2. Two possibilities for the kinds of convection cells that bring up heat from the mesosphere, and in the process cause plates of lithosphere to move across the face of the Earth. (a) Convection is confined to the weak rocks of the asthenosphere, with the mesosphere serving as a giant heating unit. (b) Convection involves the entire mesosphere, and the asthenosphere is the hot, plastic top to the convection system.

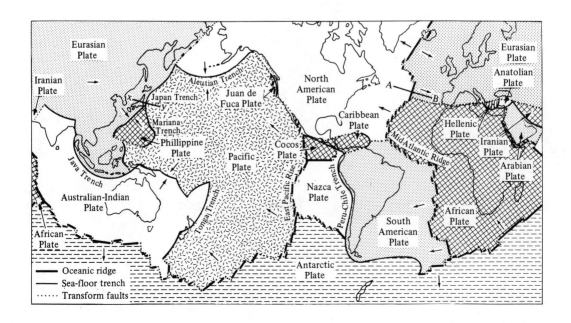

be causing the plates of the lithosphere to slide over the asthenosphere (Fig. 8.3).

Although the plates move at velocities only up to 10 cm (4 in.) per year, we can actually measure those movements. This is one of the marvels of the space age. Satellites can be used to measure the movements of lithosphere plates directly and accurately.

By extrapolation from well-supported data, we can state such things as the following: When the plates drift headlong into each other, a mountain range will be squeezed up. When a plate is cracked, and the two fragments drift apart, a new ocean basin may result. When two plates simply slide past each other, they do so along a great fracture such as the San Andreas Fault, and people living near the fracture must learn to live with earthquakes.

Igneous rock occurrences

Igneous rocks occur where they do because of plate tectonics. Most igneous rocks that are forming today, or that formed in the recent past, occur along or near margins of plates.

When a large plate splits and an ocean basin forms, the margin separating the two plates typically runs down the middle of the ocean basin. As the newly formed plates drift apart, magma rises up from the mantle and continually fills the widening rift with new igneous rock. As an example, Europe and North America were once parts of a single continent that split up about 200 million years ago; the two plates that Europe and North America now sit on are still drifting apart at a rate of about 5 cm (~2 in.) per year. The Mid-Atlantic Ridge, a giant, submarine volcanic mountain chain that marks the ancient rift, lies equidistant between them (Fig. 8.4). In a few places, such as in Iceland, the chain breaks the surface of the sea, and we can actually see, and walk on, volcanoes of that predominantly submarine volcanic mountain range. The kinds of igneous rocks found along plate margins formed through such splitting are basaltic and gabbroic rocks.

Plate margins that move toward, rather than away from, each other present a completely different scenario. Along these margins, the plate

Figure 8.3 (facing). The lithosphere is divided into six large tectonic plates and several smaller ones. The plates move continually in the directions shown by arrows. Plates have three kinds of margins: (1) spreading edges, coincident with midocean ridges (see Fig. 8.4, a section through the Mid-Atlantic Ridge, along the line A–B); (2) subduction edges, delineated by seafloor trenches, island-arcs, and volcanoes (see Figure 8.5a, a section through the subduction edge of the Pacific Plate, along the line X–Y); and (3) transform fault margins along which two plates slide past each other.

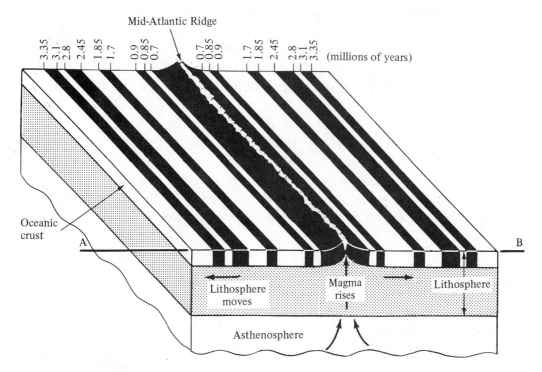

Figure 8.4. Schematic diagram of the Mid-Atlantic Ridge, along the line A–B in Figure 8.3. Magma rising from the asthenosphere forms new oceanic crust as the lithosphere is pulled apart at equal rates in both directions. The cooling oceanic crust is magnetized with a normal (i.e., present-day) north pole during epochs of normal magnetic field (black), or reversed during epochs of a reversed magnetic field (white). Other data allow the dates of magnetic reversals to be determined, and the rates of plate motions can be calculated from these dates. The approximate dates are shown, in millions of years past, for the last nine major reversals of the magnetic field.

made of the more dense rocks sinks into the asthenosphere, and the other plate, made up of less dense rocks, rides over it. The sinking process is called *subduction* (Fig. 8.5a). The subducting plate carries down water-saturated rocks that have been on the ocean floor. As these rocks become heated at depth, they melt – at least in part – and the magma so formed rises up to form volcanic edifices such as those that comprise the Cascades and the Andes. Mount St. Helens in the Cascades, Catopaxi in the Andes, Mount Fuji in Japan (Fig. 8.5b), and Mount Mayon in the Philippines are all examples of the andesitic stratovolcanoes formed along subduction margins.

The rock of which the continental crust is constructed has such a low density that continental crust cannot be subducted. Therefore, when two plates made up largely of continental crust collide, a mountain chain is formed by an accordionlike crimping and crumpling (Fig. 8.6). Examples include the Himalaya Mountains, which were formed when India

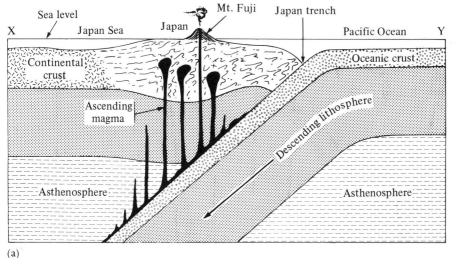

Figure 8.5. Subduction. (a) Section through the subduction edge of the Pacific Plate, along the line X–Y in Figure 8.3. Descending lithosphere, capped by wet oceanic crust, starts to melt at a depth of about 80 km. Magma rises to form steep-sided stratovolcanoes such as Mt. Fuji. (b) Mt. Fuji, Japan, is one of several andesitic stratovolcanoes that have been formed near subduction margins. (Photograph by D. R. Crandell)

collided with Asia; the Appalachians, formed when Europe and northern Africa hit North America some 400 million years ago; and also the Alps, the Caucasus, and the Urals. One result of such collisions is that both plates become greatly thickened along their colliding edges, and that thickening causes deep burial of the crust, as well as a raising-up of the crust. In turn, the deep burial may initiate melting of the continental crust involved, and the magma thus formed can rise upward and consolidate to form such rocks as granite, granodiorite, and other silica-rich igneous rocks.

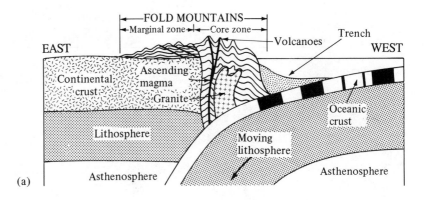

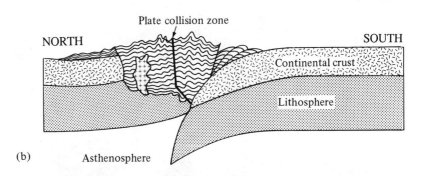

Figure 8.6. Collision of plates. **(a)** Fold mountain of the kind that make up the Andes of South America. The down-going plate of lithosphere is the Nazca Plate (see Fig. 8.3). The high Andes of Chile, Peru, and western Bolivia is the section labeled core zone; the marginal zone is eastern Bolivia and western Argentina (viewed as looking south). **(b)** When a down-going plate brings a second fragment of continental crust into collision with the first, a mountain range of the Himalayan type results. The downward-moving plate of lithosphere becomes detached, but the edge of the subducting plate is partly thrust under the stationary plate, thus causing further elevation of the mountain. **(c)** The south wall between Lhotse and Nuptse, two 8,000-m neighbors of Mt. Everest, in the Himalaya. The visible strata, which were laid down in the sea more than 90 million years ago, have been lifted to their present position as a result of the collision of the Indian subcontinent with the rest of Asia. (Photograph by Adam Stern)

Rocks and minerals in diverse environments

Figure 8.7. Migmatite. This migmatite from central Minnesota is the kind of rock that is often formed when rocks of diverse compositions are buried so deeply that some of them – typically those of granitic composition – are heated to the point that they become highly mobile while the other rocks remain less mobile and in some cases are merely metamorphosed. (Photograph by C. D. Putnam)

Metamorphic rock occurrences

Regional metamorphic rocks occur along both subduction- and collision-type plate boundaries. As noted in Chapter 4, metamorphism involves changes in form, texture, and mineral assemblages produced in pre-existing rocks as the result of increases in temperature and/or pressure and/or changes in chemical environment (Fig. 8.7).

Sedimentary and pyroclastic rocks are particularly susceptible to pronounced metamorphism. Sediments tend to accumulate along margins of continents, where continental crust is next to oceanic crust, and broad, submerged shelf areas are present. Large quantities of pyroclastic rock are ejected from andesitic volcanoes that form above margins of subduction. Consequently, sediments and pyroclastic rocks are common in areas involved in the actions accompanying subduction or collision of plates and there take on the characters that we see as folded piles of metamorphic rocks and migmatites in many mountain ranges.

Sedimentary rock occurrences

We have described several aspects of both physical and chemical weathering, transport, deposition of sediments, and the formation of sedi-

Gems, granites, and gravels

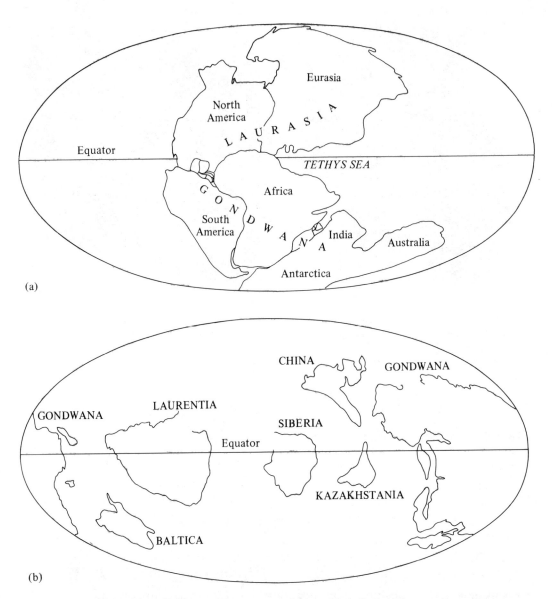

Figure 8.8. Two views of the ancient world. **(a)** Two hundred million years ago the continents of today were grouped into a single, giant continent: The northern half, consisting of North America, Europe, and Asia, is given the name Laurasia; the southern half, consisting of all the other continents and fragments, is called Gondwana. **(b)** Five hundred and forty million years ago, Gondwana dominated the map, Asia had not been assembled into a single piece, and Laurentia, the core of today's North America, straddled the equator. The place where New York now stands was then in the Southern Hemisphere, about where Rio de Janeiro is today. Open boundaries indicate that geological information is incomplete. (Adapted from Bambach, R., et al., American Scientist, vol. 62, p. 26, 1980)

mentary rocks in Chapters 4–7. Our descriptions direct attention to the fact that large quantities of minerals (e.g., calcite, halite, gypsum, goethite, and clay minerals) and of rocks (e.g., residual clays and ores, and diagenetic rocks as well as sedimentary rocks) originate on or near the Earth's surface. As mentioned, these minerals and rocks are formed as a result of such processes as oxidation, both physical and chemical transport, sedimentation, and lithification. The sedimentary rocks are derived, then, as a result of the exposure of igneous, metamorphic, and older sedimentary rocks to the elements – that is, to the atmosphere, the weather, and biological activities. And they form and occur on the interior parts as well as along the margins of plates (Plate 51).

These sedimentary and diagenetic rocks comprise nearly 75 percent of all rocks that are exposed on the Earth's surface or that occur directly below the regolith. Included are such rocks as shale, sandstone, limestone, and dolostone.

THE ROCK CYCLE

Today, plate tectonics can be observed and studied in progress, and the associations of igneous, metamorphic, and sedimentary rocks that mark plate boundaries can be seen in deeply eroded rocks as old as 2.5 billion years. Nonetheless, the question is still open as to exactly how long plate tectonics has been operating. We can only say that at no time in the past did the map of the world look like today's map, nor will it ever look the same again (Fig. 8.8). We know this because the system requires

that continents continually break up, assemble into new pieces, break up again, and move,
that ocean basins open up and accumulate sediment around their edges, and
that the sediment be squeezed into mountain ranges as the oceans close up.

We also know that as mountain ranges have been raised up, erosion has worn them down to form low-lying plains, usually within a few tens or a hundred or so million years. Someday, it seems, even the mighty Himalaya will probably be a low-lying inland plain.

In addition, we know

that debris derived from the erosion of mountains moves downslope under the influence of gravity, is carried by rivers to the sea, is deposited, and may become sedimentary and diagenetic rocks,
that eventually those rocks will very likely be squeezed into a new mountain range, and
that the cycle of events may start all over again.

Gems, granites, and gravels

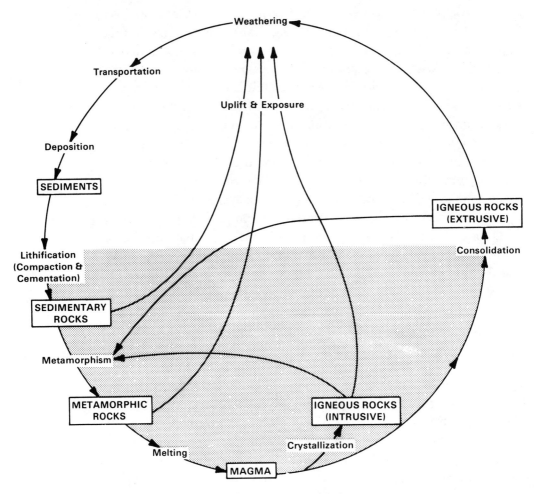

Figure 8.9. This "rock cycle" diagram shows how the main kinds of rocks and geological processes of the Earth's crust may be interrelated. Rocks and rock materials are enclosed in boxes; processes are not. Shaded and unshaded areas represent subsurface and surface domains, respectively. [From Dietrich, R. V., 1989; Stones: Their Collection, Identification, and Uses (2nd edition). Geoscience Press, Prescott, Arizona, with permission]

There is, in fact, good evidence that all of these things have happened many, many times during the Earth's history. Thus, the overall cycle of erosion, deposition, deep burial, metamorphism and melting, uplift, erosion, and so forth, has come to be called the *rock cycle* (Fig. 8.9). And its existence was recognized by James Hutton, the father of modern geology, as early as the late years of the 18th Century.

Hutton, however, did not know what drove the cycle, nor was there in his day any understanding of how old the Earth is. Thus, his statement that there is "no vestige of a beginning, no sign of an end" to the rock cycle must be said a bit differently today. We now know the age of the

Earth, and thus we also know that the Earth's rock cycle can be no older than about 4.6 billion years; therefore, there was indeed a beginning. We agree, however, that there is no sign of an end, and also that the rock cycle has continued – granted, with both interruptions and short circuits – for so long that there is no place on Earth that has not been shaped and sculpted by the rock cycle. We also feel sure that there is no large group of rocks on Earth that has not been formed as a result of, or influenced by, plate tectonics. Hence, we conclude that all minerals and rocks, ore deposits, petroleum and coal, and other resources we obtain from the Earth are direct or indirect results of processes involved in the rock cycle and plate tectonics.

EPILOGUE

This is a remarkable planet on which we live. No other planet in the solar system has been shown to have both a rocky lithosphere and plate tectonics. The Earth seems to be geologically unique, at least within this solar system.

Perhaps there is an even more important uniqueness – one that guides the destiny of our planet: Some scientists point out that the atmosphere is the way it is because living plants and animals keep it that way. It also is known that the existence of water standing on the surface of the Earth requires the temperature-ameliorating atmosphere that is present. It is suspected that water may play a vital, though still little understood, role in the motion of the Earth's plates.

Is it possible that the persistence of life requires the processes and results of plate tectonics and the rock cycle? Is it possible, too, that plate tectonics requires the existence of life?

FURTHER READING

Cloud, P., 1988, *Oasis in Space: Earth History from the Beginning.* Norton, New York, 508p.
Written especially for the layperson, this is a good treatment of the historical evolution of the Earth's crust and atmosphere, including the formation of the continents, plate tectonics, and the evolving biosphere.

Cox, A., and Hart, R. B., 1986, *Plate Tectonics: How It Works.* Blackwell Scientific, Oxford, England, 392p.
An engagingly written introductory text. The senior author, Cox, was one of the major contributors to the whole concept of plate tectonics and the proof of its correctness as supported by paleomagnetic data (some mathematics involved).

Eicher, D. C., McAlester, A. L., and Rottman, M. L., 1984, *The History of the Earth's Crust*. Prentice-Hall, Englewood Cliffs, New Jersey, 197p.
An introductory paperback text that discusses the role of plate tectonics throughout the Earth's long history.

Lovelock, J. E., 1972, Gaia as seen through the atmosphere. *Atmos Environ*, 6, 579–80.
This is the paper in which the designation Gaia – "the Greek personification of mother Earth" – was given to the hypothesis that holds that life controls as well as is controlled by the global environment, for example, the atmosphere.

Skinner, B. J., and Porter, S. C., 1987, *Physical Geology*. Wiley, New York, 750p.
This fine introductory textbook provides a good introduction to the "plate tectonics revolution."

Uyeda, S., 1978, *The New View of the Earth* (translation by M. Ohnuki). W. H. Freeman, San Francisco, 217p.
A nonmathematical, easy-to-read introduction to plate tectonics. Can be read in conjunction with the book by Cox and Hart.

Appendix 1: Chemical symbols and the periodic table

Each chemical element has been given a one- or two-letter symbol that has been accepted on a worldwide basis. Symbols for the elements that are in the formulas of relatively common minerals are given in Table A1.1.

All of the chemical elements are generally presented in what is known as the periodic table of the elements (Fig. A1.1). This table gives the chemical elements in sequence according to their atomic numbers. The periodic table is organized into horizontal tiers called periods and vertical columns called groups.

Elements in the same period differ from each other in a systematic manner from one end to the other of their tier. From left to right, their outer shells are progressively filled with additional electrons until the outer shell characteristic of the period is filled to its capacity, thus giving the appropriate noble-gas element.

Elements constituting a group are often termed *congeners* because they have similar physical and chemical properties. For elements in the "A" groups, the outermost shells contain electrons equal to the Roman-number group designation – for example, each of the elements in group IIA (the elements with atomic numbers 4, 12, 20, 38, 56, and 88) has two electrons in its outermost shell.

Ions for elements in groups IA, IIA, and IIIA are positively charged *cations;* ions for elements in VA, VIA, and VIIA are negatively charged *anions.* The groups are frequently named and may be briefly described as follows:

Group IA, the alkali metals – soft, light metals; most strongly electropositive, highly reactive.

Group IIA, the alkaline-earth metals – harder, heavier metals; strongly electropositive; reactive; easily form oxides, hydroxides, carbonates, sulfates, etc.

Group IIIA, the boron or aluminum group – boron (B) has properties intermediate between the metals and nonmetals, sometimes leading to the designation *metalloid;* aluminum (Al) and its other cogeners are metals; form oxides.

Group IVA, the carbon–silicon group – silicon (Si) and germanium (Ge) are metalloids; tin (Sn) and lead (Pb) are metals; carbon (C) plays a role in the organic world similar to that played by silica (Si) in the inorganic world.

Group VA, the nitrogen or phosphorus group – a mixed group in which nitrogen (N) and phosphorus (P) are nonmetals, arsenic (As) and antimony (Sb) are metalloids, and bismuth (Bi) is generally considered to be a metal.

Appendix 1

Table A1.1. *Chemical elements and complexes in the formulas of relatively common minerals*[a]

Chemical elements				Anionic groups	
Ag	Silver	Mo	Molybdenum	AsO_4	Arsenate
Al	Aluminum	N	Nitrogen	BO_3	Borate
As	Arsenic[b]	Na	Sodium	CO_3	Carbonate
Au	Gold	Nb	Columbium	CrO_4	Chromate
B	Boron	Ni	Nickel	MoO_4	Molybdate
Ba	Barium	O	Oxygen[b]	NO_3	Nitrate
Be	Beryllium	P	Phosphorus	OH	Hydroxyl
C	Carbon	Pb	Lead	PO_4	Phosphate
Ca	Calcium	Pt	Platinum	SO_4	Sulfate
Ce	Cerium	S	Sulfur[b]	SiO_4	
Cl	Chlorine[b]	Si	Silica	Si_2O_7	
Co	Cobalt	Sn	Tin	SiO_3	Silicates
Cr	Chromium	Sr	Strontium	Si_4O_{11}	
Cu	Copper	Ta	Tantalum	Si_4O_{10}	
F	Fluorine[b]	Th	Thorium	UO_2	Uraninate
Fe	Iron	Ti	Titanium	VO_4	Vanadate
H	Hydrogen	U	Uranium	WO_4	Tungstate
Hg	Mercury	V	Vanadium		
K	Potassium	W	Tungsten		
Li	Lithium	Zn	Zinc		
Mg	Magnesium	Zr	Zirconium		
Mn	Manganese				

[a] To facilitate reference, the elements are listed alphabetically according to their symbols, and the anionic groups are listed alphabetically by the first letter of each group.
[b] These elements typically act as anions, and the resulting compounds are called, for example, oxides, fluorides, and sulfides.

Group VIA, the oxygen group – oxygen (O), sulfur (S), and selenium (Se) are nonmetals; tellurium (Te) and polonium (Po) are generally considered to be metalloids.

Group VIIA, the halogen group – nonmetallic; most strongly electronegative; highly reactive.

Group VIIIA, the noble gases – chemically inert; form very few compounds.

Groups IB through VIIB and VIII, the transition metals – each of these groups, which constitute the central portions of the long periods of the table, has one of eight rather complex sets of chemical properties.

The lanthanides, along with lanthanum (La) and yttrium (Y), and sometimes scandium (Sc), are rather widely referred to as *rare earths* or rare-earth metals.

The actinides and actinium (Ac) are radioactive elements sometimes referred to as the *uranium metals*. The transuranium man-made elements are included.

To the present, 91 elements have been found to occur naturally on Earth. Technetium (Tc), promethium (Pm), and the transuranium elements, except for plutonium (Pu), have not been found to occur naturally on Earth.

key

Atomic number
Symbol
Atomic Weight

IA	IIA	IIIB	IVB	VB	VIB	VIIB	GROUP VIII	GROUP VIII	GROUP VIII	IB	IIB	IIIA	IVA	VA	VIA	VIIA	VIIIA
1 **H** 1.008																	2 **He** 4.003
3 **Li** 6.939	4 **Be** 9.012											5 **B** 10.811	6 **C** 12.011	7 **N** 14.007	8 **O** 15.999	9 **F** 18.998	10 **Ne** 20.18
11 **Na** 22.99	12 **Mg** 24.31											13 **Al** 26.982	14 **Si** 28.086	15 **P** 30.974	16 **S** 32.07	17 **Cl** 35.453	18 **Ar** 39.948
19 **K** 39.102	20 **Ca** 40.08	21 **Sc** 44.956	22 **Ti** 47.90	23 **V** 50.942	24 **Cr** 51.996	25 **Mn** 54.938	26 **Fe** 55.847	27 **Co** 58.933	28 **Ni** 58.71	29 **Cu** 63.546	30 **Zn** 65.38	31 **Ga** 69.735	32 **Ge** 72.59	33 **As** 74.922	34 **Se** 78.96	35 **Br** 79.904	36 **Kr** 83.80
37 **Rb** 85.47	38 **Sr** 87.62	39 **Y** 88.906	40 **Zr** 91.22	41 **Nb** 92.906	42 **Mo** 95.94	43 **Tc** 98.906	44 **Ru** 101.07	45 **Rh** 102.905	46 **Pd** 106.4	47 **Ag** 107.868	48 **Cd** 112.41	49 **In** 114.82	50 **Sn** 118.69	51 **Sb** 121.75	52 **Te** 127.60	53 **I** 126.904	54 **Xe** 131.30
55 **Cs** 132.905	56 **Ba** 137.33	57 **La** 138.91	72 **Hf** 178.49	73 **Ta** 180.948	74 **W** 183.85	75 **Re** 186.2	76 **Os** 190.2	77 **Ir** 192.2	78 **Pt** 195.09	79 **Au** 196.967	80 **Hg** 200.59	81 **Tl** 204.37	82 **Pb** 207.19	83 **Bi** 208.980	84 **Po** (~210)	85 **At** (210)	86 **Rn** (222)
87 **Fr** (223)	88 **Ra** 226	89 **Ac** (227)															

58 **Ce** 140.12	59 **Pr** 140.908	60 **Nd** 144.24	61 **Pm** (145)	62 **Sm** 150.35	63 **Eu** 151.96	64 **Gd** 157.25	65 **Tb** 158.925	66 **Dy** 162.50	67 **Ho** 164.930	68 **Er** 167.26	69 **Tm** 168.934	70 **Yb** 173.04	71 **Lu** 174.97
90 **Th** 232.038	91 **Pa** 231.036	92 **U** 238.03	93 **Np** 237.05	94 **Pu** (244)	95 **Am** (243)	96 **Cm** (247)	97 **Bk** (247)	98 **Cf** (251)	99 **Es** (254)	100 **Fm** (257)	101 **Md** (258)	102 **No** (259)	103 **Lr** (260)

Figure A1.1. Periodic table of the elements.

FURTHER READING

CRC Handbook of Chemistry and Physics. CRC Press, Cleveland.

This annual publication, which has gone through more than 69 editions, is the standard reference for chemical and physical properties of elements and both organic and inorganic compounds. It lists several chemical data and physical constants in an easily utilized format.

Appendix 2: Identification of the common rock-forming minerals

The minerals included in the two tables in this appendix are those that must be identified in order to name the common rocks. In fact, the coverage in the tables is limited to the common rock-forming varieties of the tabulated minerals; for example, the characteristics of rose quartz and orange feldspar (var. "sunstone") are not included.

The method outlined for making mineral identifications in hand specimens can be used to identify essentially any mineral, common or rare. To identify less common minerals it would only be necessary to utilize more extensive determinative tables than those given here (see "Further reading" at the end of this appendix).

Most of the common minerals can be identified relatively easily once one becomes familiar with their physical properties and how to subject them to appropriate, typically rather simple, tests. In the tables, the minerals are divided into two main groups: those that appear metallic (Table A2.1) and those that appear nonmetallic (Table A2.2). The minerals in each of these groups are arranged in order of their Mohs hardness values (see Table 2.1), and each mineral is color-coded. Each mineral name is followed by a simple and generally reliable test that will help confirm its identification.

The following procedure is suggested for using the tables:

1. Note whether the mineral appears metallic or nonmetallic.
2. Note the mineral's color.
3. Determine the mineral's approximate hardness.
4. On the basis of entries in the "Remarks" column, determine other properties that may be definitive.

For example, if the mineral (1) is nonmetallic, (2) is white, and (3) has a hardness slightly greater than that of one's fingernail (i.e., $>2\frac{1}{2}$), it is likely to be calcite, anhydrite, or dolomite. Next, use the "Remarks" column: If the mineral effervesces briskly in dilute HCl (hydrochloric acid : water $\cong$ 10 : 90), it is calcite; if it effervesces only slowly in dilute HCl, generally only after being powdered, it is dolomite; if it is dissolved, without effervescence, in concentrated HCl, it is anhydrite.

One caution must be kept in mind: There is always the possibility that the mineral being checked is not included in these two tables. If so, perhaps the

Table A2.1. *Determinative table for the most common metallic minerals*

gray/ black	red/ orange	yellow/ green	Hardness	Name	Remarks
×			2½–3	Galena	Lead-gray color; 3 perfect cleavages at right angles to each other; decomposed by nitric acid (HNO_3).
×	×	×	3½–4	Sphalerite	Resembles resin in reflected light; crystal faces are commonly curved; orange fluorescence in ultraviolet light; soluble in HCl.
		×	3½–4	Chalcopyrite	Brassy-appearing; iridescent tarnish common; soluble in nitric acid (HNO_3).
×		×	4–5½	Limonite	Yellow-brown when powdered; common as coatings on weathered surfaces of iron-bearing minerals; may appear nonmetallic.
×	×		5–6	Hematite	Red-brown when powdered; common in reddish-colored rocks; may appear nonmetallic.
×			5½–6½	Magnetite	Eight-sided crystals and rounded grains are common; strongly magnetic.
		×	6–6½	Pyrite	Brassy-appearing; cubes and 12-sided crystals are common; powdered, is soluble in nitric acid (HNO_3); sometimes called "fool's gold."

mineral can be identified by using more inclusive determinative tables. Until one has gained experience using such tables, however, it would be wise to have a professional mineralogist or petrologist check the tentative identification.

FURTHER READING

Dietrich, R. V., and Wicander, E. R., 1983, *Minerals, Rocks, and Fossils*. Wiley, New York, 212p.
  This book, written for people interested in becoming acquainted with and possibly collecting minerals, rocks, and fossils, includes a table similar to that given in this appendix, but covering many more minerals.

Pough, F. H., 1976, *A Field Guide to Rocks and Minerals* (4th edition). Houghton-Mifflin, Boston, 317p.
  This is probably the best of several English-language books written for nonprofessional mineral collectors so far as its correctness and authoritative coverage. It is rather well illustrated, with color plates.

Table A2.2. Determinative table for the common nonmetallic rock-forming minerals

colorless/ white	gray/ black	red/ orange	yellow/ brown	green/blue/ purple	Hardness	Name	Remarks
X	X		X	X	1–3	Clays	Sticks to tongue; smells clayey when breathed on; group name for several hydrous aluminum silicate minerals that occur most commonly as weathering products.
X					2	Gypsum	One good cleavage; soluble in HCl; common in evaporite sequences.
X	X				2½	Halite	Salty taste; cleaves into cubes; common in evaporite deposits.
				X	2–3	Chlorite	Cleaves into thin inelastic plates; common in schists and gneisses.
X				X	2–4	Mica: muscovite biotite	Cleaves to thin, elastic plates.
X	X	X			3	Calcite	Effervesces briskly in dilute HCl; cleaves into blocks with diamond-shaped faces; exhibits double refraction.
X	X		X	X	3–5½	Zeolites	Group of hydrous aluminum silicate minerals that are common in amygdules in volcanic rocks; many occur in bundles or are fibrous; some exhibit orange or yellow-green fluorescence.
X	X				3½	Anhydrite	Soluble in hot HCl; three good cleavages at right angles to each other, but typically granular to massive in rocks.
X	X				3½–4	Dolomite	Effervesces slowly with dilute HCl when powdered; cleaves into blocks with diamond-shaped faces.

						Hardness	Mineral	Description
	X					4–5½	Limonite	Yellow-brown when powdered; a common weathering product of iron-bearing minerals.
X	X					5–5½	Glass	This nonmineral constituent of many volcanic rocks resembles quartz; glassy appearance; conchoidal fracture; see Chapter 4.
	X		X			5–6	Hematite	Earthy-appearing; red-brown when powdered.
	X			X		5–7	Amphibole	Two cleavages at ~55° and 125°; typically lathlike.
	X			X		5–7	Pyroxene	Two cleavages at 90°; typically blocklike.
X	X			X		5½–6	Nepheline	This is the most common feldspathoid; may appear glassy or greasy; decomposed by HCl; commonly weathers out, leaving associated feldspar in relief; does not occur with quartz.
					X	6	Epidote	Pistachio to bilious green.
	X					6–6½	Alkali feldspar	Two cleavages at 90°, flesh to salmon color is common.
X			X			6–6½	Plagioclase feldspar	Two cleavages at nearly right angles; parallel lines may be seen on better cleavage.
			X		X	6–7½	Garnet	Glassy to dull appearance; equidimensional crystals are common; irregular breakage.
X			X		X	6½	Chalcedony (agate, etc.)	Microcrystalline; waxy luster; agate is color-banded; jasper is reddish.
X					X	6½–7	Olivine	Glassy; apple green; conchoidal fracture; does not occur with quartz.
X	X		X		X	7	Quartz	Glassy appearance; conchoidal fracture.
	X		X		X	7	Tourmaline	Lengthwise striated prisms with cross sections that resemble spherical triangles; electrically charged on heating and cooling.

Appendix 3: The identification of rocks

Anyone can learn to identify the common rocks. This appendix has been set up to help one develop appropriate procedures and thus gain the experience that is necessary to identify rocks correctly in hand specimen.

Information is presented for both the major rock families (igneous, sedimentary, and metamorphic rocks) and the minor families (pyroclastic, diagenetic, and migmatitic rocks), as discussed in Chapter 4. The rocks included constitute a very large percentage of all of the rocks of the Earth's crust.

The first need is to decide to which major or minor family a given rock belongs. For the most part, this is relatively straightforward.

>Most igneous rocks consist largely of interlocking grains of silicate minerals.
>
>Detrital sedimentary rocks consist of clasts such as sand grains and clay particles; chemical and biochemical sedimentary rocks, like igneous rocks, consist largely or wholly of interlocking grains, but the minerals are carbonates, sulfates, and salts.
>
>Many metamorphic rocks are foliated; those that are not typically exhibit obviously distorted features or contain minerals not frequently found in other rocks.
>
>Pyroclastic rocks are much like detrital sedimentary rocks, but consist of fragments derived directly from volcanic eruptions.
>
>Diagenetic rocks, other than the coals, resemble chemical and biochemical sedimentary rocks and can be distinguished from them only because of their mineralogical compositions or their textures, and/or by microscopic studies.
>
>Migmatites are clearly mixtures of two or more components: one a foliated metamorphic rock, the other appearing to be igneous.

The few exceptions to these generalities will become readily apparent to anyone involved with rocks, particularly as rocks are identified in the field.

IGNEOUS ROCKS

Igneous rocks, except for glasses, are identified on the basis of their grain size and mineral content (Table A3.1). Many petrologists consider the component minerals to be of four kinds:

>*Specific minerals* are those required by the definition of the rocks.
>
>*Varietal minerals* are generally present, and their names usually are used as adjectives in a rock's designation.

The identification of rocks

Table A3.1. *The common igneous rocks*

	Phanerites[a]	Aphanite[b] equivalents
	Syenite	Trachyte
	Granite	Rhyolite
	Granodiorite	Dacite
	Diorite	Andesite
	Gabbro	Basalt
	Peridotite	

(Diagram: PERCENT BY VOLUME — showing Biotite Mica, Na-rich Alkali feldspars, Plagioclase feldspar, Quartz, Hornblende, Pyroxene, Olivine, Ca-rich, Ab₅₀An₅₀; with axis "Darkness and Specific Gravity Increase"; each division represents 10 percent.)

Notes: Adjacent rocks listed in the vertical columns grade into one another. Except for peridotite, for which olivine is the specific mineral, the specific minerals are those that are predominantly white on the diagram.

Glasses are called obsidian if they have chemical compositions similar to that of granite or granodiorite, tachylyte if their chemistry is that of a gabbro (or basalt).

[a] Phanerites are rocks the essential minerals of which can be identified with the naked eye or with only the aid of a 10× hand lens.

[b] Aphanites are rocks in which more than 50% of the constituents cannot be distinguished with the aid of a 10× hand lens. Most aphanites contain some visible grains and may be named correctly on the basis of the identities and relative quantities of those grains.

Source: R. V. Dietrich, *Stones, Their Collection, Identification, and Uses* (2nd edition). Geoscience Press, Prescott, Arizona, 1989, with permission.

Accessory minerals are typically present in small amounts, but usually are not mentioned except in complete descriptions.

Alteration products may or may not occur and, when present, may or may not be alluded to in the designation – when they are, their names are usually followed by the suffix *-ized* (e.g., epidotized) to direct attention to the fact that they are thought to have been formed as a result of something that happened to the rock after it had become solidified from its parent magma.

Table A3.2. *Example of igneous rock composition*

Mineral	Composition vol. %)	Category
Quartz	22.0	Specific minerals
Alkali feldspar	46.0	
Plagioclase feldspar	13.0	
Biotite	15.0	Varietal mineral
Apatite	00.5	Accessory minerals
Zircon	00.3	
Others (e.g., magnetite)	00.2	
Chlorite	03.0	Alteration mineral

An example of igneous rock composition is given in Table A3.2. Quartz and the two feldspars are the specific minerals; biotite is a varietal mineral; the other minerals are accessories, except for chlorite, which is present as an alteration of biotite. The rock would be called a partially chloritized biotite granite.

Another kind of igneous rock that is not in the table, but may be encountered, comprises a group of which each rock is made up of 90 percent or more of a single mineral. These are (each with its predominant mineral) syenite (alkali feldspar), anorthosite (plagioclase feldspar, typically one of intermediate composition), pyroxenite (a pyroxene), dunite (olivine), magnetitite (magnetite), and chromitite (chromite).

Yet another group of relatively common igneous rocks are called *porphyries*. These rocks consist of relatively large crystals, called *phenocrysts*, surrounded by smaller grains called the *groundmass*. Phenocrysts in most porphyries are predominantly, if not wholly, crystals of one or more of the specific minerals. The groundmass may be either phaneritic or aphanitic. The name porphyry is added to the appropriate rock name. Two examples are syenite porphyry, which would typically have alkali feldspar phenocrysts in a syenite-composition groundmass, and trachyte porphyry, which would have alkali feldspar phenocrysts in a trachyte groundmass.

SEDIMENTARY ROCKS

Sedimentary rocks are usually subdivided into detrital rocks and chemical and biochemical rocks. Detrital rocks are named on the basis of the size of their predominant constituent clasts (Table A3.3). Chemical and biochemical rocks are named on the basis of their mineralogical compositions (Table A3.4).

Although it is frequently said that the presence of fossils is limited to sedimentary rocks, fossils also occur in pyroclastic and diagenetic rocks and have been recognized in a few metamorphic rocks.

Table A3.3. *Detrital clasts, sediments, and sedimentary rocks*

Clast	Diameter (mm)	Loose sediment	Sedimentary rock
Boulder	More than 256	Gravel or rubble[a]	Conglomerate or sedimentary breccia[a]
Cobble	64 to 256		
Pebble	2 to 64		
Sand	$1/16$ to 2	Sand	Sandstone[b]
Silt	$1/256$ to $1/16$	Silt	Siltstone
Clay[c]	Less than $1/256$	Clay	Shale or mudstone[d]

[a] If particles are rounded, the loose sediment is called gravel, and the lithified equivalent is a conglomerate; if they are angular, the loose sediment is rubble, and the rock is a sedimentary breccia.
[b] *Arkose* is a sandstone that consists of 25% or more feldspar grains; *greywacke* is a highly impure, typically gray-to-green-colored sandstone.
[c] Clay, used in this sense, refers to a particle size, not (except coincidentally) to the minerals called clays.
[d] Shale splits into thin layers and flakes parallel to the sedimentary layering. Mudstone has the same general composition, but it commonly includes a noteworthy percentage of silt-sized fragments and typically constitutes relatively thick, blocky layers.

Table A3.4. *Chemical and biochemical sedimentary rocks*

Rock name	Chief mineral constituent
Limestone[a]	Calcite
Travertine[b]	Calcite
Chalk	Calcite and/or aragonite
Rock gypsum	Gypsum
Rock anhydrite	Anhydrite
Rock salt	Halite
Chert[c]	Chalcedonic quartz
Dolostone[d]	Dolomite

[a] Some limestones are clastic and then are sometimes given special designations, such as calcirudite (if predominantly gravel-sized fragments), calcarenite (if sand-sized), and calcilutite (if silt- or clay-sized).
[b] Travertine is the name given to cave deposits.
[c] Much chert is diagenetic.
[d] Some geologists call these rocks dolomite; many, if not all, dolostones are diagenetic.

METAMORPHIC ROCKS

Metamorphic rocks are subdivided into two groups: the foliated metamorphic rocks and the nonfoliated metamorphic rocks. The foliated rocks are named on the basis of the perfection of the foliation and/or the grain size of the constituent grains (Table A3.5).

Appendix 3

Table A3.5. *The common foliated metamorphic rocks*

Name	Features
Gneiss	Imperfect foliation or banding; granular minerals – typically quartz or the feldspars – predominate.
Amphibolite	Poorly to well foliated; green to nearly black amphiboles plus or minus off-white plagioclase predominate. (If quartz is present in notable percentages, the rock is called amphibole gneiss or schist, as is appropriate.)
Schist	Well-developed, closely spaced foliation; platy minerals – commonly one or more of the micas or a chlorite – appear to predominate.
Phyllite	Intermediate in grain size between schists and slates; glossy luster; commonly appears corrugated.
Slate	Homogeneous-appearing; so fine-grained that constituent minerals cannot be distinguished under a hand lens; can be readily split into thin slabs, the planes of which may transect the original bedding planes.

The most common nonfoliated metamorphic rocks are quartzite and marble. Unfortunately, some quartz-cemented sandstones have also been called quartzite. In fact, in hand specimens, some metamorphic quartzites and quartz-cemented sandstones are indistinguishable. If a rock is known to be one or the other because of its geological occurrence or because of microscopic studies, it is best to call it either *metamorphic quartzite* (or *metaquartzite*) or *sedimentary quartzite*, as is appropriate. There is a similar problem with some marble specimens: Some coarsely recrystallized limestones and dolostones closely resemble some marbles. Furthermore, in the marketplace, any carbonate rock that will take a polish is usually called a marble. We think that the term *marble* should be restricted for application to metamorphic rocks that consist chiefly of calcite and/or dolomite.

One other convention dealing with the naming of metamorphic rocks employs the prefix *meta-*. If, for example, a conglomerate is known to have been metamorphosed, it is properly termed a metaconglomerate.

PYROCLASTIC ROCKS

Pyroclastic rocks are named on the same basis as detrital sedimentary rocks – that is, on the sizes of their constituent fragments (Table A3.6).

It also is noteworthy that some geologists use the adjectives "vitric" (for glass), "lithic" (for rock), "crystal" (for mineral), and "mixed" to further describe pyroclastic debris and rocks. These, of course, describe the makeup of the predominant constituents. Thus, for example, a lapilli tuff consisting largely of fragments of glass would be termed a vitric lapilli tuff.

Table A3.6. *Pyroclastic debris and rock*

Size of fragments	Unconsolidated debris	Rocks[a]
>64 mm	Blocks, bombs, and spatter[b]	Pyroclastic breccia, agglomerate, and agglutinate
2–64 mm	Lapilli	Lapillite or lapillistone
<2 mm	Ash	Tuff

[a]Many pyroclastic rocks are mixtures of diverse-sized fragments; so, for example, lapilli tuff is relatively common.

[b]*Blocks*, the chief constituents of volcanic breccia, are angular fragments representing, for the most part, pieces of shattered lava. *Bombs* are formed when clots of partly fluid lava solidify while flying through the air. *Splatter* and its rock equivalent, agglutinate, consist of fragments that represent "squashed" clots of lava that were still partly fluid when they hit the ground.

DIAGENETIC ROCKS

Diagenetic rocks are frequently classified with their sedimentary precursors. Except for members of the coal group (Table A3.7), these rocks have not been given special names. A few petrologists, however, distinguish these rocks by using such adjectives as "recrystallized" or "replaced," or even the term "diagenetic" (or "diagenic") itself. Along with the coals, most dolostones and many cherts and coarsely crystalline limestones are examples of rocks that owe their current identity largely to diagenesis.

MIGMATITES

Migmatites are mixtures of metamorphic rocks – typically dark-colored amphibolite or biotite-rich gneiss – and a light-colored granitic rock. Several score names have been applied to these composite rocks. In some cases, up to fifteen different names have been applied to the same or quite similar exposures. This leads to the suggestion that for everyone except migmatite experts it is probably best to call these rocks just migmatites.

FURTHER READING

Dietrich, R. V., and Skinner, B. J., 1979, *Rocks and Rock Minerals*. Wiley, New York, 319p.
  This book gives the names and descriptions of all common and several less common rocks. Available in both English and as a German translation, it is widely used by students, professionals, and interested laypersons.

Appendix 3

Table A3.7. *The coal group of rocks*

Name	Description
Peat[a]	Light tan to dark brown; spongy mass of partly decomposed plant remains that commonly resembles plug tobacco; dried, it tends to crumble and will burn readily.
Lignite	Chocolate brown to nearly black; dull to pitchlike luster; dense and compact to earthy and fragile; tends to shrink on drying; burns with a smoky yellow flame, and frequently emits a strong odor.
Bituminous coal	Dark gray to velvet black; typically banded, with alternating layers of dull, pitchy, or vitreous lusters; brittle, and commonly exhibits jointing at about right angles to its banding, and thus it tends to break into blocks; burns readily, with smoky yellow flame, emitting an oily odor; this is the coal usually marketed as soft coal; it will coke.
Anthracite coal	Dark gray to jet black; nearly homogeneous; conchoidal fracture; some of it exhibits iridescent, peacock colors on broken surfaces; burns with a nearly smokeless and odorless flame; this is the coal usually marketed as hard coal; it is *not* metamorphic as some people suggest.

[a] Peat is not a rock in any sense of the word. It is, interestingly, almost always reported as a mineral resource, no matter what its use.

Dietrich, R. V., and Wicander, E. R., 1983, *Minerals, Rocks, and Fossils: A Self-teaching Guide*. Wiley, New York, 212p.
This book provides a good introduction for those who want to learn how to identify rocks (and also the common minerals and fossils).

Tennissen, A. C., 1983, *Nature of Earth Materials* (2nd edition). Prentice-Hall, Englewood Cliffs, New Jersey, 415p.
Textbook on minerals and rocks, their naming and classification.

Index

Plate numbers, where applicable, are given in italics. All plates appear on unnumbered pages between pp. 54 and 55.

acid rain, 123
adobe, 133
Agassiz, Louis, 93
agate, *Pl. 15; see also* chalcedony
agglomerate, 163
agglutinate, 163
albite, 49; *see also* feldspar
alexandrite, *see* chrysoberyl
Alhambra Palace, Granada, Spain, 134, *Pl. 47*
Alps, 68, 90, 138, *Pl. 50*
alum, 16
amazonstone, *see* feldspar
amber, 28, 60
amethyst, 28, 49, *Pls. 11, 15; see also* quartz
amorphous solids, 15
amphibole, 58, 157
amphibolite, 122, 162
Andes, 68, 138, 144
   Cotopaxi volcano, 142
andesite, 122, 159
Angkor Wat, Cambodia, 85
anglesite, 50
anhedral grain, 52–3
anhydrite, 58, 156, 161
anion, *see* ion
anorthosite, 106
anthracite, *see* coal
apatite, *Pl. 4*
   hardness, 26
   nonmetallic mineral deposits, 111, 112
   in rock phosphate, 108
Appalachian Mountains, 68, 70, 73

Appian Way, Rome, 121
aquamarine, *see* beryl
arkose, 161, *Pl. 44*
aragonite, 28, 67, 120
   origin of name, 50
argentite, 103
ash, 163
asphalt, 60
Aspidin, Joseph, 131
asthenosphere, 138, 139, 140, 142
Atlas Mountains, 68
atom, 15, 39
atomic theory, 35
Australia
   Ayers Rock, 72, *Pl. 28*
   zircon placers in, 89
australites, 73
azurite, 111, *Pl. 14*

babefphite, 49
Babylon, 96, 133
balancing rocks, 72, *Pl. 27*
barite, 47, *Pl. 39*
   deposits, 112, 114
   origin of name, 49
Barlow, William, 22
Bartholinus, Erasmus, 31, 32
Bartholomé, Galapagos, 72
basalt, 121, 122, 128, 159
   vesicular, 8
bastnasite, 103
bauxite, 62, 109, 110
bed load, 90; *see also* clast
beryl, 28, 47, 102, *Pl. 11*

Index

bertrandite, 102
Berzelius, Jöns Jakob, 37
biomass, 115
biotite, *see* mica
bismuthinite, 102
bituminous coal, *see* coal
Blake, Eli Whitney, 125
block, pyroclastic, 163
bloodstone, *see* chalcedony
Blowing Rock, North Carolina, 72
Bohr, Niels, 36
bomb
  pyroclastic, 163
  volcanic, 63
bonds, 40–1
borax, 47, 112, 114
bornite, 102
boulder, 161, *Pl. 42; see also* clast
Boyle, Robert, 36
Bragg, William Henry, 24, *Pl. 9*
Bragg, William Lawrence, 24, *Pl. 9*
brannerite, 103
breccia, 161, 163
brick, 132
brine, 112
Building stone, 120, 121, *Pl. 44*
Burma, gem placers in, 89
Butte, Montana, 100

caddis fly, *Pl. 41*
cairngorm, 49, *Pl. 15; see also* quartz
calcite, 14, 17, 36, 43, 44, 46, 47, 58, 67, 112, 120, 156
  hardness, 26
  Iceland spar, 17, 29, 31, 32
  in lapis lazuli, 28
  origin of name, 49
  in soil, 83
  weathering, 79
caliche, 83, 85, 86
Carlsbad Caverns, New Mexico, 73
carnallite, 112
carnelian, *see* chalcedony
carnotite, 47, 103
Cascade volcanoes, 138, 142
cassiterite, 100, 103, *Pl. 4*
cation, *see* ion
cat's-eye, *see* chrysoberyl
celestite, 49
cement, 130–1

Central African Republic, diamond placers in, 89
cerite, 103
cerussite, 11, *Pl. 36*
chalcedony, 27, 49, 157, *Pl. 11; see also* quartz
  agate, 28, 49, *Pl. 15*
chalcocite, 102, 111
chalcopyrite, 100, 102, 155
Chaldeans, 96, *Pl. 32*
chalk, 66, 122, 161, *Pl. 23*
chemical elements, 151, 152, 153
  trace, 46
chemical symbols, 37, 151, 152
  historic, 38
chert, 67, 68, 161, 163
Chimney Rock, Nebraska, 72
Chladni, Ernst Florenz Friedrich, 33
chlorite, 58, 156
chromite, 102, 106, 107
chromium, discovery of, 36
chrysoberyl, 28
chrysolite, *Pl. 11*
chrysoprase, *see* chalcedony
cinnabar, 102
citrine, 28, 49; *see also* quartz
clarain, *see* coal
clast, 65, 87, 88, 161
  sizes of, 161
clay, 36, 58, 156, 161
  kaolinite, 47, 80, 83
  montmorillonite, 83
  nonmetallic mineral deposits, 111
cleavage, 17, 30
cliffordite, 50
coal, 59, 60, 67, 99, 115, 116
  formation, 117
  types, 60, 164
cobble, 87, 161, *Pl. 42; see also* clast
coesite, 44, 45, 75
colemanite, 112, 114
Colosseum, Rome, 122
columbite, 103
comets, 71
concrete, 129, 130
conglomerate, 122, 161
constancy of interfacial angles, law of, 18
constant composition, law of, 35
coordination number, 42, 43
copper, 96, *Pls. 32, 33*
coquina, 66
core, of Earth, 139, 140
Cornwall, England, 105

Index

corundum, 27, 49, *Pls. 4, 14*
   hardness, 26
Coster, Dirk, 36
covalent bond, 41
cristobalite, 44, 45
crocoite, 36, 47, *Pl. 13*
crushed stone, 124
crystal, 13
   classes, 22
   physics, 25
   shapes, 52–3
   structure, 2, 3
   systems, 19, 20
crystalline solids, 15, 25
crystalline state, 13
crystallography, 10, 15

dacite, 159
Dalton, John, 35, 38
Davy, Humphry, 1
de Nehou, Louis Lucas, 136
density, 29
Devil's Tower, Wyoming, 72, *Pl. 26*
diabase, 122, 128
diagenesis, 67, 68, 163; *see also* rock, diagenetic
diamond, 1–4, *Pls. 1, 2, 10, 11*
   Atomic packing, 1, 2
   hardness, 26
   Hope Diamond, 27, *Pl. 10*
   polymorphism, 44
   synthetic, *Pl. 3*
Diamond Head, Hawaii, 72, *Pl. 25*
diaspore, 102, 109
dietrichite, 50
dimension stone, 121, 122
diposide, *see* pyroxene
diorite, 122, 159
dolerite, 122
dolomite, 58, 67, 102, 156
   Kona, Michigan, 127
   weathering, 79
dolostone, 67, 122, 128, 161, 163
Dundas District, Tasmania, *Pl. 13*
durain, *see* coal
duricrust, 122
dust, 9, 10, 89
Dwars River, Transvaal, South Africa, 106

East Pacific Rise, 104
Eddystone Lighthouse, 131
Einstein, Albert, 115

elbaite, *see* tourmaline
electron, 38, 39
emerald, *Pl. 11; see also* beryl
enargite, 102
epidote, 58, 157, *Pl. 12*
Etta Mine, Keystone, South Dakota, 108
eucryptite, 102
euhedral grain, 52–3
exinite, *see* maceral

Federov, Evgraf Stepanovich, 22
feldspar, 28, 36, 44, 108, 157, *Pl. 12*
   alkali, 58, 157
   cleavage, 30
   hardness, 26
   nonmetallic mineral deposits, 111
   orthoclase, 28, 49
   plagioclase, 8, 49, 56, 58, 117, *Pl. 5*
   potassium, 80
feldspathoid, 58
fission, nuclear, 115
fleischerite, 50
fluorite, 46, 47, *Pl. 39*
   deposits, 112, 114
   hardness, 26
foliation, 70, 71, 162
"fool's gold," *see* pyrite
Forbidden City, Beijing, *Pl. 46*
forsterite, 47; *see also* olivine
fossil, 66, 67
   fuel, 115
Fountain of Freedom, 126
Friedrich, Walter, 24
frost wedging, 79
fulgurite, 58, 62
fusain, *see* coal
fusion, nuclear, 115

gabbro, 122, 159, *Pls. 30, 44*
galena, 47, 102, 155
gangue, 99
garnet, 28, 58, 88, 157, *Pls. 5, 11*
   fracture, 30
garnierite, 102
gas, 115
   natural, 99
Gaspé Copper Mine, Québec, 106
gems, 27–9, *Pls. 10, 11*
   simulated vs. synthetic, 28–9
geology, definition, 10
geothermal energy, 115
geothermal gradient, 139

Index

Giant's Causeway, Northern Ireland, 72
gibbsite, 47, 102, 109
Gibraltar, Rock of, 72
glacial drift, 92, 93
glacial erratics, 93
    Big Rock, Beaver Island, Michigan, 94
glacial till, 92, 93
glass, 58, 135, 157; *see also* fulgurite; obsidian; pele's hair; pumice; tachylyte
glauberite, 112
gneiss, 70, 71, 162, *Pl. 5*
goethite, 81, 102
    as iron ore, 108
    in laterite, 85, 109
    in soil, 83
gold, 37, 47, 96, 100, 102, *Pl. 16*
    in lapis lazuli, 28
    in placers, 88, 111
Gondwana, 146
Gore Mountain, New York, *Pl. 5*
gossan, 110
grain shapes, 52
Grand Canyon, 72, *Pl. 51*
granite, 122, 128, 159, *Pls. 20, 44*
    Barre, Vermont, 127
    Brandywine, Delaware, 127
    Elberton, Georgia, 127
    Massachusetts, 124
    Mount Airy, North Carolina, 125, 127
    Pikes Peak, Colorado, 126
    Stony Creek, Connecticut, 127
    weathering, 78
granodiorite, 56, 159, *Pl. 12*
graphite
    atomic packing, 1, 2
    polymorphism, 44
gravel, 86, 87, 129, 161
greenstone, 122
greywacke, 161
Grindstone City, Michigan, 65
groundmass, 160
gypsum, 43, 58, 122, 156, 161
    antiquity of name, 49
    deposits, 113
    hardness, 26
    nonmetallic ore, 108
    plaster, 134
    weathering, 79

hafnium, discovery of, 36
Half Dome, *see* Yosemite Valley
halite, 24, 36, 43, 44, 58, 156
    cleavage, 30
    crystal model, *Pl. 9*
    deposits, 111, 112, 113
    hardness, 25–6, 154
Häuy, René Just, 18
hematite, 47, 58, 155, 157
    as iron ore, 102, 108
hemimorphite, 47
Hessel, Johann Friedrich Christian, 22
hiddenite, *see* spodomene
High Sierra, California, 138
Himalaya, 68, 142, 144, 146
Hooke, Robert, 15, 16, 17
hornblende, *Pl. 5; see also* amphibole
Hutton, James, 68, 148
Huygens, Christiaan, 17, 29
Hwang Ho (Yellow River), China, 90
hydrothermal alteration, 105
hydrothermal solution, 100, 104, 105, 106
    deposits formed by, 101
    submarine, 104

ice, 13, *Pl. 7*
Iceland spar, *see* calcite
igneous rocks, *see* rock, igneous
ignimbrite, 122
ilmenite, 43, 44, 103, 111
impactites, 71, 73
indochinites, 73
inertinite, *see* maceral
International Mineralogical Association, 50, 51
ionic bond, 41
ions, 39
    anions vs. cations, 39, 151
    complex, 40
    groups, 40
    polymerization, 46, 48
    radii, 39, 42
    substitution, 44
iron formation, banded, *Pl. 35*

jade, *see* jadeite; nephrite
jadeite, 28

kaolinite, *see* clay
Khorat Plateau, Thailand, 14
Kilauea, Hawaii, *Pls. 19, 22*
kimberlite, 3, 4
Knipping, Paul, 24
Kola Peninsula, USSR, 104
kunzite, *see* spodomene
kyanite, *Pl. 4*

Index

lapilli, 163
lapis lazuli, 27, 28, *Pl. 11*
larvikite, 123
laterite, 83, 85, 109, 110, 122, *Pl. 31*
Laurasia, 146
Laurentia, 146
lautarite, 47
lava, *see* magma
lazurite, 28
lechatelierite, 44, 45
lignite, *see* coal
limestone, 161
    building stone, 121, 128, *Pl. 44*
    dimension stone, 122
    fossiliferous, 66
    Key Largo, 127
    Louisiana, 127
    nonmetallic ore, 108
    Salem, Indiana, 124
    Tennessee, 122, 127
    weathering, *Pl. 29*
limonite, 58, 155, 156
linnacite, 102
Linnaeus, Carolus, 49
liptinite, *see* maceral
lithification, 67
lithosphere, 138, 139, 140, 141, 142
loess, 90
    in Yenan, China, 91
Long Island, New York, 93
Lyell, Charles, 92

maceral, 57, 59, 60
magma, 64
    lava, 64, *Pl. 19*
    residual, 107
magnesite, 45, 46, 102, 112
magnetite, 37, 58, 88, 102, 155
    vanadiferous, 103
magnetism, 37
malachite, 111, *Pl. 14*
manganese nodules, 108, 109
manganite, 44
marble, 70, 79, 122, 162
    Georgia, 124
    weathering, 78
    Yule, Colorado, 124
Martha's Vineyard, Massachusetts, 93
mesosphere, 139, 140
metacinnabar, 102
metallic bond, 41

metamorphism, 69, 70, 75; *see also* rock, metamorphic
meteorite, 8, 71, 73, 74, 75
Mexico, University Library, *Pl. 43*
miagyrite, 103
mica, 30, 156
    biotite, 56, 58, 81, *Pl. 12*
    cleavage, 30
    muscovite, 58
microscope, 33
Midatlantic Ridge, 141, 142
migmatite, *see* rock, migmatite
Minas Gerais, Brazil, 108
mineral
    accessory, 58, 159, 160
    assemblages, 56
    categories, 47
    chemistry, 5
    classification, 46
    collection, 5
    determinative tables, 155, 156, 157
    extraterrestrial, 7
    formation, 51
    formulas, 43, 44
    number, 50
    occurrences, 53
    optical properties, 31
    ore, *see* ore mineral
    rock-forming, 57, 58
    specimens, 6–7
mineral charcoal, *see* coal
mineral deposit, 98
    biochemical, 114
    banded-iron formation, *Pl. 35*
    by chemical sedimentation, 108
    disseminated, 105
    evaporite, 113
    hydrothermal, 101, 114
    in igneous rocks, 105
    by magmatic segregation, 106, 107
    metallic, 101
    nonmetallic, 111, 112
    ore, *see* ore deposit
    residual, 109
    by weathering, 109
mineralogy, 9
Mohs, Friedrich, 19, 26
    hardness scale, 26, 154
moldavites, 73
molybdenite, 102
monazite, 102, 103

Index

Mont-Saint-Michel, France, 123
moondust, 71
moonstone, *see* feldspar
moraines, 93
morganite, *see* beryl
mortar, 130
mountains
 fold, 144
 volcanic, 142
Mount Cook, New Zealand, 92
Mount Everest, Tibet, 144
Mount Fuji, Japan, 142, 143
Mount Mayon, Philippines, 142
Mount Pelée, Martinique, 63
Mount St. Helens, Washington, 2, 63, 142, *Pl. 25*
mud, 86
mudstone, 161

Namibia
 diamond placers, 88
 Namib Desert, 86
natural gas, 99
nepheline, 64
 properties, 157
nephrite, 28
neutron, 38, 39
nitratine, 47
Noachian Flood, 92
Nobel Prize, 23, 24, 36
Notre Dame Cathedral, Paris, *Pl. 48*
nuclear energy, 115

obsidian, 58, 59, 122, *Pl. 17*
oil, 99, 115
Old Man of the Mountains, New Hampshire, 72
olivine, 8, 9, 28, 58, 81
 properties, 157
opal, 28, 60, 61
 precious, 61, *Pl. 18*
optical properties, 31–2
ore deposit, 98, 99; *see also* mineral deposit
ore mineral, 99
 metallic, 102
 nonmetallic, 112
orthoclase, *see* feldspar

pahoehoe, *Pl. 22*
Parker, James, 131
paste, *see* gems
pearl, 28

peat, *see* coal
pebble, 161; *see also* clast
pegmatite, 107, 108
pele's hair, 59, 63, *Pl. 21*
pentlandite, 102
Percé Rock, Québec, 72
peridotite, 159
periodic table, 153
petrology, 55, 57, 62
phenocryst, 160
phillipinites, 73
phyllite, 70, 162
placer, 88, 89, 111
plaster, 134, *Pl. 47*
plate tectonics, 138, 141, 144
platinum, 89, 100, 102, 111
Pliny, 49, 135
Plymouth Rock, Massachusetts, 72
polarized light, 32, 33
pollucite, 102
polymerization, 46–8
polymorphism, 44, 45, 67
porphyry, 64, 122, 160
Portland cement, *see* cement
pozzuolana, *see* cement
Precambrian shield, 68
 Canadian, granite from, 127
proton, 38, 39
Proust, J. L., 35
puddingstone, 122
Pugwash Mine, Nova Scotia, 113
pumice, 59
pyragyrite, 111
Pyramid, Great, of Cheops, 120, 121, 137
Pyrenees, 68
pyrite, 21, 58, 81, 155, *Pls. 4, 16*
 antiquity of name, 49
 in lapis lazuli, 28
pyroclastic debris, 163
pyroclastic rocks, *see* rock, pyroclastic
pyrochlore, 103
pyrolusite, 102, 108
pyroxene, 8, 9, 58, 157
 diopside, 47, *Pl. 4*

quartz, 13, 17, 36, 45, 47, 49, 56, 58, 67, 83, 88, 157, *Pls. 4, 8, 15*
 deposits, 111
 fracture, 30
 hardness, 26
quartzite, 70, 162

Index

Rainbow Bridge, Utah, 73
rare earths, 102–3, 152
rational indices, law of, 19
reflection, 31
refraction, 31
    double, 31, 32
regolith, 77, 81, 87
    movement of, 85, 86
    wind transport of, 89, 90
resources
    energy, 99, 115
    metallic, 97
    mineral, 99
    nonmetallic, 98
rhodocrosite, *Pl. 6*
rhyolite, 122, 159
riprap, 125, *Pl. 45*
rock, 55, *Pl. 5*
    classification, 61
    components, 57
    cycle, 147–9
    definition, 56, 57
    diagenetic, 59, 62, 64, 67, 122, 163–4
    igneous, 58, 62, 63, 64, 122, 141, 158–60
    metamorphic, 58, 62, 68, 70, 71, 122, 145, 161–2
    migmatite, 62, 68, 70, 71, 122, 127, 145, 163, *Pl. 24*
    origins, 62
    pryoclastic, 62, 63, 64, 122, 162–3
    sedimentary, 58, 62, 64, 65, 67, 122, 145, 160–1
rock crystal, *see* quartz
rock phosphate, *see* apatite
rock salt, 161
rock wool, 128
Rocky Mountains, 68
Romé de l'Isle, Jean Baptiste Louis, 18
Röntgen, Wilhelm Conrad, 23
rooseveltite, 50
rose quartz, *see* quartz
Rose Window, Notre Dame Cathedral, Paris, *Pl. 48*
Rotorua, New Zealand, 104
ruby, 26, 27, 49, *Pl. 11; see also* corundum
rutile, 103, 111

Saint Paul's Cathedral, London, 131
Saint Peter's Church, Vatican City, 122
Saint Pierre, Martinique, *see* Mont Pelée
salt domes, 114

sand, 86, 87, 129, 161, *Pl. 4*
    windblown, 91
sandstone, 122, 127, 128, 161, *Pl. 44*
Santa Rita, New Mexico, 101
sapphire, 26, 27, 49, *Pl. 11; see also* corundum
sard, *see* chalcedony
Saylor, David, 131
scheelite, 47, 103
schist, 70, 122, 162
Schoenflies, Arthur Moritz, 22
schorl, *see* tourmaline
seafloor chimney, *Pl. 34*
secondary enrichment, 110
sediment
    biochemical and chemical, 66
    clastic, 65, 87
sedimentary rocks, *see* rock, sedimentary
serpentinite, 122
Shark's Bay, Western Australia, *Pl. 40*
shale, 161
siderite, 102
silica polymorphism, 44–5
silt, 161
siltstone, 161
silver, 96, 103
Sistine Chapel, Vatican, 134
skinnerite, 50
slate, 70, 122, 162
Smeaton, John, 131
smithsonite, 111, *Pl. 37*
smoky quartz, 49, *Pl. 15; see also* quartz
soapstone, 27, 122
soil, 77, 81
    horizons, 82, 85
    minerals in, 83
    orders, 83
    profile, 82
    types in United States, 84
Sorby, Henry Clifton, 33
South Africa
    diamond placers in, 89
    stratigraphic relations, 4
space groups, 22
spatter, pyroclastic, 163
specific gravity, 29
speleothem, 65, 73
sperrylite, 102
sphalerite, 24, 44, 103, 155
    crystal model, *Pl. 9*
sphene, 49
Sphinx, 121
spinel, 28, *Pl. 11*

Index

spodumene, 28, 102, 108
spreading edge, *see* tectonic plates
Spruce Pine, North Carolina, 108
Sri Lanka, gem placers in, 89
stalactites and stalagmites, *see* speleothem
Steno, Nicolaus, 17, 18, 31
stibnite, 102
stishovite, 44, 45
Stone Age, 96
Stonehenge, England, 123
Stone Mountain, Georgia, 72
Stopes, Marie, 59
strata, 65, 67
stratified drift, 93
   Brier Hill, New York, 94
stromatolites, 120, *Pl. 40*
strontianite, 49
stylolite, 127
subduction, *see* tectonic plates
subhedral grain, 52–3
Sugar Loaf, Brazil, 72
sulfur, 112, *Pl. 38*
sunstone, 157; *see also* feldspar
suspended load, 90; *see also* clast
swedenborgite, 47
Sydney Opera House, Australia, 132
syenite, 122, 159
sylvite, 112, 113
symmetry, 19, 20–2

tachylyte, 58, 59
taconite, 98
talc, hardness, 26
tantalite, 103
tanzanite, *see* zoisite
tectonic plates, 140–3
tektite, 14, 71, 73, 74
   fracture, 14
Tennant, Smithson, 1
tetrahedrite, 102, 147
thenardite, 112, 113
thin section (of rock), 33, *Pl. 12*
tiger's-eye, *see* quartz
tile, 132, *Pl. 46*
tombarthite, 50
topaz, *Pl. 11*
   gem, 27, 28
   hardness, 26
tourmaline, 21, 28, 46, 157
   elbaite, 108
   formula, 44
   schorl, 44

trachyte, 122, 159
transform fault, *see* tectonic plates
traprock, 122, 128; *see also* basalt
travertine, 122, 127, 161
tridymite, 44, 45
troilite, 75
trona, 112, 113
tuff, 64, 122, 163
turquoise, 28, 47, *Pl. 11*

ulexite, 112
uniformitarianism, principle of, 62
United Nations Building, New York, 131
Ural Mountains, 68
   platinum placers in, 89
uranium deposits, 116
uraninite, 103

valence, 40
van der Waals bond, 41
Vauquelin, Louis Nicolas, 36
vein, 105
Versailles, Hall of Mirrors, 136
vesuvianite, 47
veszelyite, *Pl. 6*
vitrain, *see* coal
vitrinite, *see* maceral
volcanism, 63
   kimberitic, 3
volcano, 2
   Kilauea, Hawaii, *Pls. 4, 22*
   stratovolcano, 143
   see also Mount St. Helens
Von Hevesy, György, 36
Von Laue, Max, 24

weathering, 77
   chemical, 78, 79, 81, *Pls. 29, 30*
   mechanical, 79, 81
   physical, 80
Weiss, Christian Samuel, 19
whewellite, 47
White, Canvass, 131
White Cliffs, Dover, 66
Widmanstätten texture, 75
wind transport, 89, 90
witherite, 49
wolframite, 103

X-rays, 23
   diffraction, 24, 25

Index

Yellowstone National Park, 104
   Grotto Geyser, 105
Yeman region, China, 92
Yosemite Valley, 63, 72, *Pl. 20*
Young, Thomas, 33

Zaire, diamond placers in, 89
zincite, 49
zircon, 28, 101, 103, 111
zoisite, 28
zunyite, 50